Constructing Local Environmental Agendas

Local environments are becoming increasingly unsustainable – environmentally, socially and economically. Increasing wealth in the West creates more pollution, congestion and degradation of species and their habitats, often at the environmental and social expense of the South.

The local element of Agenda 21, agreed by the nation states present at the Rio Earth Summit in 1992, is the most ambitious international attempt both to address environmental problems at the local level and encourage full local democratic participation in policy making.

Constructing Local Environmental Agendas draws on original contributions from the UK, Europe, Australia, Sri Lanka and Pakistan, to argue that there *is* scope for local areas to improve their environments, provided that local people are involved. International case studies throughout the book demonstrate the importance of respect for indigenous knowledge, the need for all groups – especially those usually exluded through disadvantage – to be involved in the decision-making process, and the need to remove layers of bureaucracy from policy making.

Constructing Local Environmental Agendas provides an invaluable insight into the experiences of parallel projects across the world, particularly in the UK and the rest of Europe, Australia, Sri Lanka and Pakistan.

Susan Buckingham-Hatfield is a Lecturer in Geography at Brunel University, London and **Susan Percy** is a Senior Lecturer in Environmental Planning and Policy at South Bank University, London.

Constructing Local Environmental Agendas

People, places and participation

Edited by Susan Buckingham-Hatfield
and Susan Percy

London and New York

First published 1999
by Routledge
11 New Fetter Lane, London EC4P 4EE

Simultaneously published in the USA and Canada
by Routledge
29 West 35th Street, New York, NY 10001

Typeset in Galliard by M Rules
Printed and bound in Great Britain by Mackays of Chatham PLC,
Chatham, Kent

British Library Cataloguing in Publication Data
A catalogue record for this book is available from the British Library

Library of Congress Cataloging in Publication Data

Constructing Local Environmental Agendas : people, places and
participation / edited by Susan Buckingham-Hatfield and Susan Percy.
p. cm.
Includes bibliographical references and index.
1. Environmental policy. 2. Local government.
I. Buckingham-Hatfield, Susan. II. Percy, Susan.
GE170.C6428 1999
363.7'0056—dc21 98-26088 CIP AC

ISBN 0 415 17063 x (hbk)
ISBN 0 415 20118 7 (pbk)

This book is dedicated to Judith Matthews, 1952–1998

Contents

Figures and tables

Figures

Tables

Contributors

Valerie A. Brown has held senior positions in research, teaching and administration in environmental science and in public health. She is currently Foundation Professor of Environmental Health, School of Applied and Environmental Sciences, University of Western Sydney, Hawkesbury and Co-ordinator of the Local Sustainability Project, working on council/community monitoring of local sustainability, and on the role of local knowledge in the local application of global programmes.

Valerie is author and editor of over 16 books and monographs on local-to-global health and environment issues, including *Acting Globally: the Changing Role of Local Government in Integrated Environmental Management* (DEST, 1994), *Risks and Opportunities: Managing Environmental Conflict and Change* (Earthscan, 1995) and *Landcare Languages: Talking to Each Other about Living with the Land* (DPIE, 1996). She is a past member of the National Health and Medical Research Council and the CSIRO Advisory Council, and a current member of the Foreign Affairs and Trade Non-Government Consultative Forum on Environmental Issues.

Susan Buckingham-Hatfield is Lecturer in Geography at Brunel University, London. She edited *Environmental Planning and Sustainability* (John Wiley, 1996) with Bob Evans, and has published a number of papers on women's participation in the LA21 process. Her main research interests are in women's access to environmental decision making and in barriers to effective public participation, and she is currently working with a number of local authorities on these issues.

Raff Carmen is Lecturer in Adult Education at the Centre for Higher and Adult Education at Manchester University, UK. He is author of *Autonomous Development – Humanising the Landscape* (ZED, 1996), and has published extensively on adult education, literacy and human-scale, culturally embedded, ethical development.

Michael Clark teaches Geography, Environmental Management, Development Studies and Waste Management at the Department of Environmental Management, University of Central Lancashire, Preston, where he is also Environmental Programmes Co-ordinator and admissions tutor. He is a

Member of the Lancashire Environment Forum, a former member of the Town and Country Planning Association Council and sustainable development Study Group and currently sits on the Manchester 2020 Sustainable City Region project steering group. Michael has published widely on waste management and is joint editor of *The Role of EIA in the Planning Process* (Mansell, 1988) and *Waste Location: Spatial Aspects of Waste Management, Hazards and Disposal* (Routledge, 1992).

Bob Evans is Head of Geography at South Bank University, London. He has worked as a town planner in the public, private and voluntary sectors, and is a member of the Policy Council of the Town and Country Planning Association. He has published books and articles on land use planning and environmental policy including *Experts and Environmental Planning* (Avebury, 1995), *Environmental Planning and Sustainability* (Wiley, 1996 with Susan Buckingham-Hatfield) and *Town Planning into the 21st Century* (Routledge, 1997, with Andrew Blowers). He is co-founder and co-editor (with Julian Agyeman) of the *International Journal of Local Environment.*

Claire Freeman is a senior lecturer with the School of Resource and Environmental Planning at Massey University, New Zealand, and previously lectured at Leeds Metropolitan University. The work on children's participation in Local Agenda 21 was part of a research project on Local Agenda 21 undertaken by the Centre of Urban Development and Environmental Management (CUDEM) at Leeds Metropolitan University. Other areas in which she is engaged in research include environmental planning focusing on planning and the natural environment, the conflict between development and conservation of natural areas and community participation in planning processes. Claire has also worked as a lecturer in South Africa and as a planner with the Urban Wildlife Trust.

Jane Knightsbridge-Randall is a writer and consultant and is involved on research and development projects with voluntary groups, local authorities and international organisations. She is youth consultant to the World Wide Fund for Nature UK and has developed a multi-media approach, is involved in writing policy documents and was responsible for developing and evaluating their implementation. Her Doctorate (Cranfield University, School of Management) is an analysis of social policy.

Walter Leal Filho is the co-ordinator for the Study Group on Environmental Education in Europe and undertakes research on aspects of environmental education, awareness and information. He recently completed a project looking at Agenda 21 initiatives in Europe. Until recently he was Professor of Environmental Education at Luneburg University, Germany, but is presently based at the Technical University of Hamburg-Harburg in Germany, having just started a project investigating the use of new information technologies for environmental learning.

Marek Lubelski is working with Luton Borough Council and the World Wide Fund for Nature UK, co-ordinating a LA21 project that is developing links with Peshawar, North West Frontier Province, Pakistan. He has a background in grass-roots adult and higher education, community action and research for sustainable futures.

Judith Matthews was, until her untimely death in October 1998, Principal Lecturer in Geography, Department of Geographical Sciences, University of Plymouth, where she taught Social Geography and aspects of European Environmental Planning. Her research interests were in the area of the relationships between the social processes that incorporate experience of place into identities, and the processes of participation in urban planning. Most recently, these concerns informed her research in Australia, when she was visiting fellow in the Department of Urban and Regional Planning at Curtin University, Perth and her direct involvement with LA21 processes in Devon.

Alan Netherwood has a background in the non-governmental sector and has worked with the British Conservation Trust for Volunteers. He is currently working with Community Service Volunteers (CSV) Wales on their Environment 2000 programme.

Jane Parker was a researcher in Environmental Planning at Cheltenham and Gloucester College of Higher Education where she worked on an ESRC-funded project looking at the policy processes of LA21. She now works for the Consumers' Association.

Alan Patterson worked as a mining surveyor for the National Coal Board in north Nottinghamshire before taking a degree in Environmental Sciences at the University of East Anglia. He then worked on the ESRC Future of Welfare research programme at Bath University, leaving to take an MA in Urban and Regional Studies at Sussex University. From there he went to Reading University to carry out doctoral research on the reorganisation of state involvement in the provision of water services in England and Wales. He currently lectures in the Department of Geography and Earth Sciences at Brunel University, London, where he is researching the political economy of public sector restructuring.

Susan Percy is Senior Lecturer in Environmental Planning and Policy at South Bank University, London. She has worked in local government and is a member of the Royal Town Planning Institute. Susan's main research interests are in LA21, environmental policy and planning and rural land management issues. She has published a number of papers in these areas and is currently a researcher on the National Going for Green Pilot Sustainable Communities Project. Susan is deputy editor of the journal *Planning Practice and Research*.

Paul Selman is Professor of Environmental Planning at Cheltenham and Gloucester College of Higher Education. Previously he worked in local government, and taught at the Universities of Central England and Stirling. Paul's books include *Environmental Planning* (1992) and *Local Sustainability* (1996), and he is editor of the journal *Landscape Research*. His recent research has been in the fields of landscape ecological planning, forestry and sustainable development.

Kate S. Theobald has worked in an administrative capacity in a number of local education authorities and has been involved in implementing youth training schemes. She has a degree in Applied Social Sciences from Kingston University after which she took up her present post as Research Officer in the Department of Geography and Earth Sciences at Brunel University. She has conducted research into the impact of public sector restructuring, in particular compulsory competitive tendering, on employment practices and on the local environment. Current projects include institutional and community responses to urban regeneration in west London. Kate is also completing her PhD on local government restructuring and environmental practices.

Ken Webster is Senior Education Officer – community and governance at the World Wide Fund for Nature UK. His background is in teaching, community education and community development education. Before joining WWF he established a federal approach to local community education which engaged a wide range of local providers in co-ordinating their non-formal education programmes. In 1993 he was appointed to the highly active WWF-UK Education Department, specifically to establish the new Community Education Unit, which was intended to develop WWF (UK)'s response to LA21.

Anoja Wickramasinghe is Professor of Geography at the University of Peradinaya, Sri Lanka and Co-ordinator of the Collaborative Regional Research Network in South Asia. Her research focus has been on environment, forestry, rural development, indigenous knowledge, common property management, community development and women. She is the author of several books in these areas: *Deforestation, Woman and Forestry* (International Books, 1994); *People and the Forest: Management of the Adam's Peak Wilderness* (Sri Lanka Forest Department, 1995); *Land and Forestry: Women's Local Resource-based Occupations for Sustainable Survival in South Asia* (CORRENSA, 1997). Anoja is also a trainer on participatory planning, gender and analysis and environmental management.

Acknowledgements

We would like to acknowledge the role of the Planning and Environment Research Group of the RGS-IBG, under whose auspices the editors convened the session 'Constructing Environmental Agendas' which acted as the spur to this book. Particular thanks go to the discussant of that session, Julian Agyeman and other panellists not contributing to this book: Oz Osbourne of the West Devon Environmental Network and Stella Whittaker. We would also like to thank the presenters in that session and our contributing authors, not all of whom overlap.

The authors and publisher would like to thank the following publisher for kindly giving permission for the use of copyright material:

Kogan Page Publishers for Figure 14.1 from McLaren, D., Bullock, S. and Yousuf, N. (1998) *Tomorrow's World: Britain's Share in a Sustainable Future*, London, Earthscan Publications Limited.

1 Keys to a sustainable environment

Education, community development and local democracy

Susan Buckingham-Hatfield and Susan Percy

Introduction

As we began editing the contributions to this book, world leaders were meeting in New York to evaluate the progress of commitments made in the 1992 United Nations Commission on Environment and Development and to agree a way forward in addressing the 'world's escalating environmental crisis' (Lean, 1997). Local authorities world wide had submitted their locally negotiated proposals for an environmentally sustainable environment (LA21s), the culmination of an astonishing amount of commitment both in terms of hours spent in meetings, consultations and plan preparation and in paper documents produced.

As it happens, no agreements were reached in New York for a number of reasons, including the failure of the North to transfer sufficient resources to the South to help combat the environmental degradation created by acute poverty on the one hand, and the drive to develop Third World economies on the other. Although the 1992 Earth Summit agreed that rich countries would raise their overseas aid to 0.7 per cent of Gross Domestic Product, the percentage given has fallen to 0.27 per cent, a decrease on the 0.34 per cent given at the time of the 1992 meeting (Schoon, 1997). Another contributory factor to the failure of Earth Summit 2 to accomplish anything was the inability of the US government to commit to any environmental targets – due primarily to the intransigence of the Republican majority in Congress which fiercely defends business interests over environmental and development concerns and resists regulation (Hams, 1997; Lean, 1997).

However, despite only a weak agreement on forests, a failure to agree on overseas development aid, an aviation fuel tax or an international initiative on water and the abandonment of the Political Declaration, the text on LA21 was preserved in the final main document, otherwise considered 'rather bland' (Hams, 1997: 18). This is something of a tribute to local municipalities and communities which have been, in some cases, making substantial efforts to identify and address environmental problems. Of course the success of LA21 is extremely varied, intranationally and world wide. Since its agreement in 1992, a number of local communities have seized the initiative to plan for a more environmentally sustainable future through participatory planning. LA21 offers at best a considerable opportunity, through community empowerment, to revolutionarily transform the way in which environmental planning takes place (Voisey *et al.*, 1996), yet at the same time it faces formidable obstacles to getting off the ground at all. Not least, the seemingly impossible task of combining the need for scientific rigour in environmental monitoring, audit and assessment, with the need democratically to involve a wide range of stakeholders in their development (Brugman, 1997).

There are a number of issues that need to be resolved in order for LA21 to proceed with any measure of success, and it is one of the aims of this book to identify these issues in order to enable environmental planning to move forward. Some of these problems need to be addressed at the local level, but many are structural issues which need to be addressed at a central government, or even transnational, level of decision making. As well as identifying problems, however, the authors of the following chapters highlight the contributions local communities have made to planning a locally sustainable environment within the global context, and these examples stand as beacons to inspire other communities, even though they may be working in very different contexts.

In this opening chapter we wish to introduce a number of obstacles which currently prevent LA21 operating to its full potential. We will also explore the extent of local planning for environmental sustainability which may, or may not, come under the LA21 rubric. We will argue that where LA21 exists, it is because of years of political activity on environmental issues by many people working outside the existing policy-making structures, and also owes much to a number of environmental initiatives which have pre-dated LA21. Whilst the book as a whole will examine just how these views are being incorporated into the LA21 process in Western countries (Australia, the UK and elsewhere in Europe) and in the South (particularly in Sri Lanka and Pakistan), in this introductory chapter we will for the most part be referring to the UK context. We make no claim for geographical representation: the UK experience is used because it has made the most extensive response to UNCED's call for local environmental agendas to be devised and submitted. Other European countries (see Filho in Chapter 3) and Australia (see Buckingham-Hatfield and Matthews, Chapter 8 and Brown, Chapter 11) are included because, although LA21 is less developed, comparable environmental initiatives are in force. This enables us to examine some of the advantages and problems that can inform LA21. One country from the South (Sri Lanka) is showcased to illustrate the particular problems facing environmental agenda setting in the developing world which are, in part, a result of North–South relations (Wickramasinghe, Chapter 10 and Lubelski and Carmen, Chapter 9). Nevertheless, the North has much to learn from non-Western communities with regard to environmental management – as partnerships such as the LA21 experiment in linking the UK and Pakistan, described in Chapter 9 by Lubelski and Carmen, demonstrate. The lack of environmental education, resources and imagination will be considered as obstacles to a more sustainable environment, but their contribution to innovative projects will also reinforce the necessity for these resources.

Public participation

> Perhaps most importantly of all we must involve everybody. The environment is not just about ozone layers and rain forests. It is about the quality of life of all of us, inside our homes and outside our front door.
>
> (John Prescott, 1997: 1)

One of the focuses of this book is the capacity of individuals and local communities to have an impact on environmental decision making and, hence, future environments, through the LA21 process. By opening up this debate in the first place there is an assumption that these individuals and communities either do, have the potential to, or should have such an impact. Selman and Parker (1997) argue that LA21 is opening up the decision making process so that, in cases where this is successful, stakeholders (representatives of various communities of interest) are integrally involved in deciding the future direction of their local communities.

The fact that LA21 has captured the imagination of the majority of the UK local authorities, and many more world wide, suggests that there was a vacuum in local governance to be filled (Marvin and Guy, 1997). Whether that vacuum has been created by the hollowing out of the local state through privatisation (Patterson and Pinch, 1995 and Patterson and Theobald, Chapter 12), a renegotiation of central–local government relations which favours centralisation of power – a phenomenon which Keith suggests is not diminishing under the new Labour Government in the UK (Keith, 1998) – or people's disaffection with the formal political process is unclear, although a combination of these is likely.

We would argue that this space has already been opened up by decades of social and environmental activity at the grass-roots and on the margins of political acceptability which has raised public awareness and tapped into people's sense of injustice and unfairness, particularly in countries in North America and Europe, as well as Australia and New Zealand. In the context of the UK, Reade suggests that in the last 30 years 'by far the greatest part of such enlightened social reform as has been achieved in Britain has been brought about by single issue campaigning groups' (1997: 87). Activity such as 'Reclaim the Streets', 'This Land is Ours' and 'Critical Mass', as well as numerous, more prosaically named anti-roads, anti-airports and anti-nuclear installations protests has eroded the conception that roads, airports, nuclear power stations or nuclear weapons are a necessary component of the way in which we live. This has created an opportunity for dialogue between official and unofficial politics which, as Dobson has argued, is critical for a sustainable environment (Dobson, 1995).

Notwithstanding Marvin and Guy's (1997) critique of 'the new localism' which, they say, eulogises the power of local communities and their governing institutions to create local environmental sustainability with little reference to different scales of political, institutional and commercial activity, we suggest that LA21 fulfils a need for unofficial/official dialogue which can move us closer to sustainable development at a range of levels.

However, this is not without its problems, as the contributors to this book make abundantly clear. Selman and Parker (1997) identify three contributors to the LA21 process: *community representatives* (of, e.g., tenants' or residents' groups); *agents of change* who are the dynamic individuals who initiate ideas and follow them through; and *ordinary people* unfamiliar with engagement in formal politics. Reliance on the first two participants raises issues of true, wide representation and accountability, whilst dependence on the *tireless innovator* also poses problems when they leave the neighbourhood or burn out from all their activity. Yet, in most LA21 discussions, it is these participants who are most active. Most LA21 programmes are least able to recruit individuals unused or unwilling to engage in formal political processes.

Whilst this is bemoaned, it is unrealistic to think of these people as apathetic on the environment. Much non-violent direct action is undertaken by individuals disaffected by the existing political system, whilst others will explicitly engage in environmental activity through other involvements (from recycling and sensitive shopping to growing their own food, teaching their children or through separate initiatives such as those run by the Women's Institute). As Buckingham-Hatfield and Matthews reveal in Chapter 8, the main reason given by individuals for not actively participating in LA21 was lack of time due to other voluntary group (often environmental) commitments.

So, LA21 offers a forum where the non-political can meet policy makers in a potentially mutually enriching environment. Those disaffected by public policy making (as McNaghten *et al.* (1995) have found most focus group respondents to be) can use the LA21 stage to express their dissatisfaction and alternative ideas, whilst policy makers can use it as a safe and legitimate forum to discuss potential developments with those outside the formal political process. However, people are only likely to do this if they feel that their engagement will make a difference. Batterbury (1998) has argued that this is a necessary use of LA21 in a paper which suggests that environmental activists should get involved in local policy making to try to effect incremental changes, even if their more controversial ideas are not accepted. Using his action research in Ealing, West London, Batterburg demonstrates how local transport planners adopted ideas for cycle routes which directly emanated from this dialogue.

Arguably, it is through the local councillor or representative that individuals would normally channel their concerns, and it appears to be members' priorities which drive particular environmental initiatives (Audit Commission, 1997). However, in the UK, members' concern with a sustainable environment is still low and their lack of commitment is cited as the largest single obstacle to LA21 by environmental co-ordinators (LGMB, 1996 in Audit Commission, 1997). One of the greatest challenges facing LA21 is its capacity to influence local government and to reinvigorate the representative democratic process. This could be measured, presumably, by greater interest by councillors and higher public attendance at council meetings.

As well as seeing LA21 as an interface between formal and informal politics, it can also be seen as an interface between the domestic/personal and the public/community spheres. Concerns brought to the discussions involve a range of issues from the health of family members to domestic consumption patterns, and these areas are similarly affected by the recommendations of published LA21s: choices concerning personal mobility, food and materials purchasing and broader consumption and waste disposal issues. This is an area which particularly concerns women in their social role as main carer and domestic worker. Arguably, however, it focuses the responsibility for cleaning up the environment on the relatively disempowered, when much wider global forces are responsible (see Lubelski and Carmen, Chapter 9).

Most commentators writing about LA21 argue a case for local government to take a strong role in the process, partly to guard against the hijacking of the process by voluntary groups who may not be representative of the community (see Clark and Netherwood, Chapter 4), and as an institution of representative democracy (Selman and Parker, 1997), and also because of their repository of knowledge and expertise and their statutory responsibilities. Nevertheless, they need to use this expertise judiciously, as it is frequently this very quality which alienates local people from any kind of political activity.

Experts and how to deal with them: skill up or dumb down?

> I think that the only person that is an expert about your life is you. You are the person living it; you are the person experiencing it. No other person can be as expert about your life, your values, your desires, your hopes than you are. And any process that tries to define the future and does not find a way to include your expertise on life is doomed to fail.
>
> (Lawrence, 1997)

Much has been written about the way that lay publics are excluded from discussions about health, safety and environment because they do not understand, or do not speak the scientific language (Evans, 1995; Irwin, 1995). In brief, non-experts or ordinary people with ordinary concerns about BSE, toxic dumping, export of veal calves, additives in food or radon leakage, have to skill up to participate in the debate, since the experts generally refuse to dumb down.

Couching the sustainable environment debate (itself a jargonistic term which many people do not fully understand) in technical jargon can, at best, be unintentionally divisive, but at worst can be deliberately used to exclude difficult lay interest or concern. Selman and Parker argue that 'LA21 at least creates one durable and respectable arena in which such discourse [i.e. lay views] could take place', (1997: 179) – although our experience suggests that groups jockeying for power in LA21 discussions, be they government or voluntary sector, continue to use expert terminology both deliberately and unintentionally (see Buckingham-Hatfield and Matthews, Chapter 8, and Evans and Percy, Chapter 13).

If experts refuse to dumb down, then the only way in which local people can negotiate on anything resembling equal terms is through education. In a local conference held in London by UNED-UK, education and training were overwhelmingly identified as necessary resources for effective public participation (UNED-UK, 1997: 17). This is particularly important for the groups the UN has identified for deliberate and systematic inclusion in environmental decision making: women, children and young people and indigenous people. Knightsbridge-Randall and Freeman each address the need for appropriate environmental education delivered in accessible language later in this book, whilst the issue of rural communities' empowerment through application of local knowledge is pursued by Wickramasinge, writing about Sri Lanka. Buckingham-Hatfield and Matthews also consider the alienating effect of the disdain of 'experts' when considering the ability of women to fully participate in environmental decision making. Education, therefore, is one critical resource which is necessary in the pursuit of sustainable environments.

Educating for sustainability and community development

A recurring theme that is highlighted in a number of the contributions in this book is the important role of environmental education and community development in the delivery of LA21. See, for example, the chapters by Evans and Percy, Webster and Filho. These contributors argue that environmental education programmes linked closely to community development provide a means for the delivery of the skills, information and understanding for developing sustainable communities and can additionally encourage local democracy. The LA21 process will need to incorporate these issues if it is to be truly representative of local communities.

In the 1990s, as Tilbury (1995) states environmental education attempts to educate people for sustainability in the long term, and takes a more clearly defined stance than previously about the problems of contemporary society: 'This form of environmental education is concerned with the integration of the contemporary disciplines of environmental and development education and requires reconciliation between environmental conservation and economic development' (1995: 197).

Environmental education is being actively developed in schools, and further and higher educational institutions, and as Knightsbridge-Randall comments in Chapter 7:

> There has never been a generation more informed about environmental issues than this one. For the nineties youth, some of the issues that they have been studying in school are publicised on worldwide TV culture and interpreted locally in ways specific to each culture and its setting. What this means is that we have a very well informed youth army with the knowledge and argument to take on these issues if they so desire.
>
> (p. 82)

However, young people are not always involved in the LA21 process and many of the participants have not had the benefit of the inclusion of social and environmental issues as part of the school curriculum when they studied. Many local authorities are proceeding with their LA21s without the key ingredients of environmental education and community development as foundations on which to build the process. This is despite the concept of sustainability becoming much more of a guiding principle in environmental education in the 1990s, and it is worth reminding ourselves of the most well-known definition of sustainable development taken from the Brundtland Report: '. . . development that meets the needs of the present without compromising the ability of future generations to meet their own needs' (WCED, 1987: 43).

sustainable development according to this definition is about balancing different sets of goals: environmental, economic, political, institutional and social. Tilbury points out that the 1980 World Conservation Strategy 'first redirected the goals of environmental education towards what it referred to as education for sustainable development' (1995: 197). Hence environmental education started drawing together social, political, cultural and economic factors into the discussions on the environmental situation. In 1991, the follow up to *World Conservation Strategy – Caring for the Earth* (IUCN/UNEP/WWF, 1991) was launched which highlighted the role of education in bringing about changes towards sustainable life styles. The document states:

> Sustainable living must be the new pattern for all levels: individuals, communities, nations and the world. To adopt the new pattern will require a significant change in attitudes and practices of many people. We will need to ensure that education programmes reflect the importance of an ethic for living sustainably.
>
> (IUCN/UNEP/WWF, 1991: 5, quoted in Tilbury, 1995)

Then came the UNCED Conference and again education was emphasised as a way of promoting and improving the capacity of people to deal with sustainable development issues. In addition, through the principles that underpin LA21, such as subsidiarity by increased democratisation and decentralisation, empowerment and capacity building, the role of community development also becomes critical to the LA21 process. Community development is concerned with developing self-help communities, bringing on board marginalised communities, enabling communities to be self-determined and to participate in decision making, broadening people's horizons and educating them to be responsible citizens.

However, for the ambitious goal of sustainable communities to be realised both educating for sustainability and community development have to come together.

> If environmental education is to respond effectively to the challenges of presenting a global perspective on environmental issues and problems . . . it must renew its philosophy, revitalise its process of understanding and construct a pedagogy that is relevant to the nature and demands of the new concept of development . . .
>
> (Williams, 1991: 1, quoted in Tilbury, 1995)

It seems that few local authorities are addressing environmental education of their communities nor the 'skilling up' of citizens through community development. Exceptions include where the Community Education Development Centre (CEDC) has been involved in facilitating an understanding of sustainable development principles at the community level. CEDC exists to promote and support the development of community education and works with various local authorities including Reading Borough Council's Neighbourhood Agenda 21 programme. CEDC is developing a 'development education' model approach to community participation in decision making and brings to communities skills, expertise, practical ideas and support of the key principles of empowerment, participation and capacity building. Whilst many UK local authorities are not incorporating education and community development into their LA21s, nor unfortunately is central government in its 'Going for Green Sustainable Communities' pilot projects. The national 'Going for Green' priorities are based on the narrow conceptualisation of the environment through the 'Green Code' (travelling sensibly, reducing waste, preventing pollution, saving energy and natural resources and looking after the local environment). Little attention nationally has been devoted to educating for sustainability, that is, the wider more embracing concept of sustainable development concerning, for example, issues such as the sense of place and poverty. Nor has 'Going for Green' nationally emphasised support for community development, and has assumed, rather naively, that communities will come together and develop ideas on how to tackle the Green Code issues without prioritising or allocating time and resources to development, within communities, of the necessary skills, knowledge and capacity to deal with these issues.

LA21 in the UK: local activity and central government constraints

Chapter 28 of Agenda 21 calls upon local authorities to develop action plans for sustainable development at the local level. It states:

> As the level of government closest to the people, local authorities play a vital role in educating, mobilising and responding to the public to promote sustainable development. By 1996, most local authorities in each country should have undertaken a consultative process with their populations and achieved a consensus on a 'Local Agenda 21' for the community.
>
> (para. 28.2)

The document goes on to state that: 'The overall objective is to improve or restructure the decision making processes so that consideration of socio-economic and environmental issues are fully integrated' (para. 28.2). Hence all UK local authorities have been encouraged to prepare plans for local sustainability under the banner of LA21 or other more locally derived names with perhaps more meaning for the local communities, such as Reading Borough Council's 'GLOBE' initiative, meaning 'Go Local on a Better Environment'. Since local authorities had this target date of the end of 1996 for LA21 completion, it was timely that in September 1997 the Local Government Management Board (LGMB) issued their Review of the First Five Years of LA21 in the UK.

The Review was accompanied by 35 case studies describing a set of best practice and innovative activities, and these two documents provide an informative package of UK LA21 progress. The Review and case studies demonstrate quite clearly that much hard work, commitment and imagination has been occurring since 1992. The material collected has brought together a vast amount of information, written in a language and style that is accessible to both the expert and general public. This accessibility is extremely important given the criticism that is often levelled at the complex rhetoric of Rio (see for further explanation Clark and Netherwood, Chapter 4 and Evans and Percy, Chapter 13).

The best practice case studies have an important contribution to make by providing information on exciting new initiatives which have resulted from determination, inspiration and hard work. The case studies cover many issues ranging from partnership projects on the implementation of environmental improvements for run-down housing estates, a Pension Fund Investment Strategy attempting to link sustainability issues with investment returns, and Eco-Audits aimed at facilitating improvements to environmental performance. As Selman states:

> Workers at the coalface of sustainable development need these success stories, and need an information network of exemplary practice as a means of gaining inspiration and practical information. They, though, would be the first to admit that published cameos of 'success stories' can sometimes flatter to deceive.
>
> (1998: 15)

Whilst Selman delivers a cautionary note on not getting too optimistic about LA21 there are other fears that are not allayed by the Review document and case studies. There is, for example, little reason to believe that generally LA21 is engaging with under-represented groups in society (such as women, disabled, ethnic minorities and young people), since as Buckingham-Hatfield and Matthews in Chapter 8 state in respect of women 'There is an observable trend that government (local as well as central) is transforming citizens into consumers, but even in the more protracted discussions on consumers, there is no acknowledgement that consumption activity may be gendered' (p. 96). Freeman in Chapter 6 comments on the importance of getting children more involved in the LA21 process 'Children constitute 23 per cent of the UK population, yet they are rarely considered and even more rarely consulted about the society in which they live' (p. 68). The Review does discuss new approaches to participation, for example, planning for real, parish maps and visioning techniques, but as Selman (1998) comments, new methods of participation can still be unrepresentative and do not automatically lead to greater democracy.

The claims in the Review that LA21 is forging new processes that are facilitating local democracy and community involvement in decision making are not substantiated by evidence, since what seems to be happening is that there is enhanced participation but not necessarily increased local democracy (see Evans and Percy, Chapter 13). This confusion between participation and local democracy is a key issue since long-term and meaningful moves towards sustainable societies require, amongst other things, a change in power-making structures and institutional arrangements that go well beyond the examples described in the Review and case studies. Another cautionary note is that whilst the Review claims that 73 per cent of local authorities in the UK are pursuing LA21s there is no clarification on what pursuing means – is it full commitment or partial?

In the Review's chapter 'Background and Milestones' it acknowledges that environmental protection and environmental policy have been common themes in local government well before the UNCED Conference and that often the LA21 process has been launched on the back of existing environmental initiatives (Percy, 1998). These existing initiatives had as their main focus traditional environmental management, not the more holistic concept of sustainability, and this transfer in to LA21 has obvious limitations in terms of scope and interpretation. Examples of issues that do not have a high profile in LA21s include: investment strategies, social services, welfare strategies and anti poverty; disappointing but perhaps not that unexpected considering the highly problematic political, social and economic decisions needed in these areas of activity.

However, the amount of LA21 activity in the UK is in many ways quite remarkable especially when it is remembered that it is a non-statutory process facilitated by local authorities which have been experiencing severe budgetary constraints over a number of years. In addition, particularly under the last Conservative Government, local authorities have been operating under a very difficult political relationship with the centre. As Evans points out 'central government in Britain has offered only limited support, and this only comparatively recently. For example, the government's "sustainable development: the UK Strategy" makes only passing reference to LA21' (1998: 197–8). With the election of a Labour Government the question as to whether central government's position has changed has to be asked. On the surface the evidence suggests increasing support for sustainable development, as demonstrated by, for example, the merging of the Departments of Environment and Transport into the Department of Environment, Transport and the Regions, and the sustainable development rhetoric from John Prescott on the need for an integrated transport system, the recent joint central and local government LA21 guidance and the Prime Ministerial exhortations on reducing air pollution at the Rio +5 Conference in June 1997 and in Kyoto in Japan in November 1997. The government's new sustainable development Strategy will also bring in guides to sustainability indicators and targets, which Michael Meacher, the Environment Minister, sees as 'drivers for action and change . . . which will encapsulate the social as well as the environmental and economic dimensions of sustainability' (DETR, 1997: 362). Another encouraging sign that a more holistic approach may be forthcoming includes the designation of Green Ministers in all government departments, supporting the work of the new Cabinet Committee on the Environment. This is all very heartening, however, there are a number of constraints and difficulties that still prevail, not least the maintenance of a strong central power base at Westminster which, as we suggested earlier, shows little sign of devolving.

First of all there is the issue of the lack of resources for LA21. Many local authorities are progressing their LA21 programmes on very limited budgets which are reviewed each financial year with no guarantee of continued funding. Tuxworth (1996) points out that in a LGMB survey most local authorities have simply added LA21 to existing officers' duties rather than created new posts. This severe resourcing problem is given no financial support by central government and the uncertainty that prevails over the financial underpinning of many LA21 programmes is 'demoralising and prevents long term planning for the initiative which will cause the process to oscillate wildly. Resources include not only financial commitments but also time and personnel allocations . . . ' (Hollins and Percy, 1995: 18). There is also a need for appropriate training of those local authority staff and members involved in LA21 initiatives which is particularly crucial given the long-term nature of LA21 as opposed to the short termism which politicians, local and national, inevitably work to.

In addition LA21 preparation is a non-statutory requirement, unlike, for example, the preparation of development plans. Evans comments in respect of this:

> LA21 emphasises 'bottom-up' approaches and is less concerned with the output of a final plan than with establishing the processes and approaches which will endure into the future. For most local authorities, however, accustomed to preparing plan documents which relate to existing professional or departmental responsibilities, for specific and usually statutory ends, this approach has caused some difficulty.
>
> (1998: 200)

Furthermore, the policy goal of sustainability is all-embracing, long term and holistic in nature and both central and local government find this difficult to deal with given the fairly rigid and compartmentalised structures that prevail. Blowers (1993, 1997) notes that environmental issues are trans-media, trans-boundary and trans-sectoral and this presents problems for policy makers and decision makers who operate within traditional professional limits and departmental boundaries.

Whilst for many local authorities the LA21 process has given them a degree of legitimatisation and purpose much needed to counter the 1980s and 1990s centralisation of control over local government's finances and responsibilities, there are always contradictions which are clearly demonstrated in recent legislation and regulations. Over the last few years there has been a flood of environmental legislation coming through from Europe which the UK has had to adopt, although sometimes reluctantly, as in the case of environmental assessment. The focus according to Ball and Bell (1997) is on issues such as the control of pollution, the control of hazardous substances and processes, minimisation and management of waste and the conservation of natural resources and protection of ecosystems. This is commendable but requires policy makers to both think and operate across departmental lines. The potential benefits of environmental regulation are also limited by the consequences of the requirement to contract out the delivery and management of traditional environmental services. Patterson and Theobald state that:

> CCT does not appear to lead to the emergence of a holistic approach to policy making but does lead to the fragmentation of local authority responsibilities, as individual departments lose direct provision of services to the private sector.
>
> (1996: 10)

Hence decisions become fragmented and are driven by cost-reduction criteria and financial efficiency.

In addition, to fulfil the potential of LA21 there will inevitably have to be a change in power relations and the restructuring of local authorities generally, with the initiation of some form of internal reorganisation and refocusing of policy and decision making priorities. This debate is taken further by Patterson and Theobald in Chapter 12.

However, it is also possible to argue that it is this very dislocation between the centre and the local which has stimulated local government, particularly in the UK where such a division is relatively recent, but also in federal systems such as the USA, Germany and Australia where local communities traditionally have greater autonomy. As Voisey *et al.* suggest in the UK:

> local authorities faced with crisis over their democratic role as manifested in low and falling voters . . . and the professions associated with them, embattled by contracting out and internal management reorganisations, have used the environment as self-defence.
>
> (1996: 47)

In the context of the UK it has been suggested that one of the main impetuses for LA21 has been the identification of one arena in which local authorities, progressively limited in autonomy through 18 years of a centralising Conservatism, could assert their responsibility and imagination. Unlike federal systems such as Australia and Germany, local government in the UK has no powers of general competence and acts under the specific direction of government (Voisey *et al.*, 1996: 44). Andrew Marr has suggested that 'green politics is the new socialism' (1997: 19) and LA21 appears to have been an opportunity for local government in the UK to establish a legitimate area with which to challenge a government to which they had been ideologically opposed.

Environmentally sustainable development outside LA21

LA21, of course, is not the first or the only local environmental initiative to combine environmental considerations with issues of equity and social justice. On an international scale, the International Council for Local Environmental Initiatives (ICLEI)'s 'Sustainable Cities Project' was established in 1990 with its flagship 'Sustainable Seattle Project' established in that year. Originally developed primarily as a public education and advocacy exercise, it is claimed 'that the project has made and continues to make a recognisable impact in the community . . . to engender debate about priorities and the connections between local trends' (Brugman, 1997: 62). It has been the example to many LA21 programmes.

Individual countries usually associated with a concern for the environment are not all at the forefront of LA21. For example, Norway has a fairly undeveloped LA21 programme, and yet 93 per cent of municipalities have co-operated with NGOs in managing local-level local environment/natural resource programmes under the Norwegian Government's Environmental Protection at the Local Level programme. By February 1995, 420 out of 435 municipalities had Environmental Protection Officers, 60 per cent of whom were permanent appointments. In 1992, Norway introduced a centrally funded Eco-municipalities programme in nine rural areas which would integrate 'environmental considerations into all policy mechanisms, local resource protection, local action on global problems and involving people and improving local governance' (Nysnas quoted in Voisey *et al.* 1996: 40). Likewise, in Germany, whilst only 10 municipalities were actively involved in developing and implementing LA21 by the end of 1995, 19 were involved in ICLEI's 'Sustainable Cities Project' and 188 were part of Germany's 'Climate Alliance' (Voisey *et al.*, 1996).

Friends of the Earth have been instrumental in developing local environmental planning in Norway, working with municipalities to create an 'Environmental Home Guard' in 1991; they have also collaborated with the Norwegian Union of Municipal Employees to explore the role of trades unions in creating a sustainable society. (Voisey *et al.*, 1996). In the UK, Friends of the Earth were instrumental in an early initiative to work with local authorities to help them become more environmentally sustainable (Friends of the Earth, 1989).

As Valerie Brown points out in Chapter 11, Australia (as elsewhere) has extensively used State of the Environment monitoring which came before LA21. Other recommendations of Agenda 21, such as strengthening community action and developing skills in negotiation and advocacy, have been present in other initiatives which predate it, such as the Australian Landcare programme and the international Healthy Cities programme initiated by the World Health Organisation. Brown notes striking similiarities of goals between the Healthy Cities programme and Agenda 21, with a framework for Healthy Cities comprising: intra/inter-generational equity in utilising natural resources, maintenance of biodiversity and natural capital, and the invocation of the precautionary principle. So whilst Agenda 21 is often considered to be a radical departure from previous environmental programmes, it needs to be seen more as an evolution from earlier initiatives which share similar goals and identify common problems.

Conclusion

We hope that the following chapters contribute towards a better understanding of the opportunities and constraints inherent in LA21 initiatives and environmental policy making. Whilst each chapter raises particular issues involving the construction of local environmental agendas in specific settings and contexts, there are nevertheless common and recurring themes that emerge as discussed earlier. We therefore hope that this book can help to inform policy makers, politicians, educators, academics and the public on some of the critical issues that can either aid or hamper LA21 progress.

Chapters 2 to 5 discuss, albeit in different ways, the themes of local democracy and civicness, the rhetoric of Rio and education for sustainability. We cannot understate the importance of these themes, and as Parker and Selman state in Chapter 2 '. . . during the 1990s, we have witnessed a shift of emphasis from "local government and the environment" to one of "local governance and sustainability" . . .' (p. 18). Filho notes the importance of environmental information as a component of local environmental initiatives, highlighting four variables that determine the extent and nature of public participation and ultimately increasing local democracy in LA21. Clark and Netherwood in Chapter 4 argue that LA21 is bound up in rhetoric, and is therefore open to contradictory interpretations, noting the heavy reliance on voluntary organisations in shaping and delivering LA21. Chapter 5 by Webster brings together these themes in a discussion from the World Wide Fund for Nature (UK)'s perspective which includes environmental education as a prerequisite for long-term sustainability, encouragement of participatory democracy and the full airing of the implications of achieving sustainability. The discussions in these contributions draw on UK, European and Canadian experiences. Chapters 6 to 8 concentrate on the need to include disadvantaged groups in LA21 initiatives. The chapters concentrate on the importance of inclusion of children, young people and women in decision making processes. For example Freeman, in Chapter 6, comments on how children in the UK have very little influence over how society is shaped even though the sustainable development movement places children at its centre. A case study of LA21 in Leeds highlights how children can be encouraged to participate in the process and have their voices listened to. Knightsbridge-Randall also emphasises the need for young people to be involved in LA21 and argues for a new vision for youth, allowing young people to speak and represent themselves in their own voice and style. Buckingham-Hatfield and Matthews in Chapter 8, in their discussion on including women draw on both British and Australian experiences and conclude that gender bias in the LA21 process will greatly influence the way in which women engage with sustainability and how it is vital for women's concerns and perspectives to be included in decision making processes.

The next part of the book concentrates on issues concerning developing countries, indigenous peoples and wider international relations. Lubelski and Carmen in Chapter 9 argue that it is the 'Northern' or 'First World' vision of the future that is currently shaping the agenda of international policy making, including

LA21. They discuss the problems associated with capitalist exploitation of many southern countries by the North and point out how LA21 offers up the opportunity to demonstrate forcibly to all countries the reality of global inequity, environmental degradation, and social and political exclusion. They argue that a major challenge for sustainability is to find ways of developing relations between the North and the South that are built on collaboration, co-operation and cross-cultural understanding, and they give examples of collaboration between the peripheries of northern and southern states which show how local people can learn from each other, unhampered by the intervention of central government. From a slightly different perspective, Wickramasinghe in Chapter 10 argues for these three ingredients and draws on the experience of LA21 in Sri Lanka. She emphasises the inequalities in development opportunities, resource distribution and how locally acceptable technology has been replaced by externally developed technologies, often with dire consequences. Wickramasinghe argues for the need to restore local initiatives and empower local communities drawing on local innovation and culture.

The final part of the book highlights the need for integrated and holistic policy making, the identification of barriers to and opportunities for progress in environmental policy making. In Chapter 11 Brown explores new strategies for using State of the Environment (SOE) reporting systems to link decisions at global, national and local scales. Brown argues how SOE monitoring can facilitate the incorporation of sustainability indicators into mainstream decision making, drawing on information from a number of countries. Several authors, Patterson and Theobald, and Evans and Percy discuss the problem of local-level institutions incorporating sustainable development policies into current policy-making structures and decision making processes. Patterson and Theobald (Chapter 12) particularly concentrate their discussion on the impact of compulsory competitive tendering within three environmental services concluding that local government in Britain is experiencing constraints on its ability to implement sustainable development goals. Evans and Percy provide a wider discussion of the opportunities that the LA21 process has afforded to local government and local communities but they also highlight a number of barriers to the process. Evans and Percy conclude that new approaches, including greater environmental education, increasing local democracy and changes to decision making structures will need to be evolved and implemented if LA21 is to make a real, long-term difference.

In our view these themes and issues are integral to the future shaping and framing of local environmental agendas and in our final chapter we take the opportunity to reflect on the importance of these themes and put forward a framework in which they will need to be located within environmental policy-making processes.

References

Audit Commission (1997) *It's a Small World: Local Government's Role as a Steward of the Environment*, London: Audit Commission.

Ball, S. and Bell, S. (1997) *Environmental Law* (4th edn) London: Blackstone Press.

Batterbury, S. (1998) 'Environmental Activism and Social Networks: Campaigning on Transport Issues in West London', paper presented to the *1998 RGS-IBG Annual Conference*, Kingston, Jan.

Blowers, A. (ed.) (1993) *Planning for a Sustainable Environment* London: Earthscan.

Blowers, A. (1997) 'Environmental planning for sustainable development' in Blowers, A. and Evans, B. (eds) *Town Planning into the 21st Century*, London: Routledge.

Brugman, J. (1997) 'Is there a method in our measurement? The use of indicators in local sustainable development planning', *Local Environment* 2(1): 59–72.

Department of the Environment (1994) *sustainable development: the UK Strategy* London: HMSO.

Dobson, A. (1995) 'No environmentalisation without democratisation', *Town and Country Planning* Dec.: 322–3.

Evans, B. (1995) *Experts and Environmental Planning*, Aldershot: Avebury.

Evans, B. (1998) 'The rhetoric of Rio and the problem of local sustainability' in Kivell, P.T., Roberts, P. and Walker, G. (eds) *Environment, Planning and Land Use*, Aldershot: Avebury.

Friends of the Earth (1989) *Environmental Charter for Local Government*, London: Friends of the Earth.

Hams, T. (1997) 'Was UNGASS all hot air?' *Local Environment News* 3(7): 16–18.

Hollins, M. and Percy, S. (1995) *From the Globe to the Local: an Interim Evaluation of the Reading Neighbourhood Local Agenda 21 Initiative*, Surrey: WWF (UK) and RBC.

Irwin, A. (1995) *Citizen Science*, London: Routledge.

IUCN/UNEP/WWF (1980) *World Conservation Strategy: Living Resources for sustainable development*, Nevada: IUCN/UNEP/WWF.

IUCN/UNEP/WWF (1991) *Caring for the Earth: A Strategy for Sustainable Living*, London: Earthscan.

Keith, M. (1998) 'Governing London?' paper presented to the *1998 RGS-IBG Annual Conference*, Kingston, Jan.

Lawrence, G. (1997) *The Way Forward. Building our Future, Sustainable Regeneration in North and East London*, London: UNED-UK.

Lean, G. (1997) 'Earth summit failure leaves planet in peril', *Independent on Sunday*, 26 June.

Local Government Management Board (1997a) *Local Agenda 21 in the UK – The First Five Years*, London: Local Government Management Board.

Local Government Management Board (1997b) *Local Agenda 21 Case Studies* London: Local Government Management Board.

McNaghten, P., Grove-White, R., Jacobs, M. and Wynne, B. (1995) 'Public Perceptions and Sustainability in Lancashire – Indicators, Institutions, Participation' report to Lancashire County Council by the Centre for the Study of Environmental Change, Lancaster University, Lancaster: CSEC.

Marr, A. (1997) 'A climate for change', *The Independent*, 25 June.

Marvin, S. and Guy, S. (1997) 'Creating myths rather than sustainability: the transition fallacies of the new localism', *Local Environment* 2(3): 299–302.

Patterson, A. and Pinch, P. (1995) '"Hollowing out" the local state: compulsory competitive tendering and the restructuring of British public sector services', *Environment and Planning* 27(9): 1437–61.

Patterson, A. and Theobald, K. (1996) 'Local Agenda 21, compulsory competitive tendering and local environmental practices', *Local Environment* 1(1): 7–20.

Percy, S. (1998) 'Real progress or optimistic hype?' *Town and Country Planning*, Jan./Feb. 67(1): 19–20.

Prescott, J. (1997) 'Saving the world involves us all', *Earth Summit II News*, June, London: UNED-UK.

Schoon, N. (1997) 'Blair gets serious on climate summit, US condemned for foot dragging', *The Independent*, 24 June.

Selman, P. (1998) 'A real local agenda for the 21st century?', *Town and Country Planning* Jan./Feb. 67(1): 15–17.

Selman, P. and Parker, J. (1997) 'Citizenship, civicness and social capital in Local Agenda 21', *Local Environment* 2(2): 171–184.

Tilbury, D. (1995) 'Environmental education for sustainability: defining the new focus ofenvironmental education in the 1990s', *Environmental Education Research* 1(2): 195–212.

Tuxworth, B. (1996) 'From environment to sustainability: surveys and analysis of Local Agenda 21 process development in UK local authorities', *Local Environment* 1(3): 277–97.

UNED-UK (1997) *Building our Future, sustainable development in North and East London*, London: UNED-UK.

United Nations Commission on Environment and Development (1992) *United Nations Conference on Environment and Development Agenda 21*, Geneva: UNCED.

Voisey H., Beueurman, C., Sverdrup, L. A. and O'Riordan, T. (1996) 'The political significance of Local Agenda 21: the early stages of some European experience', *Local Environment* 1(1): 33–49.

Williams, R. (1991) 'Teacher education survey: environmental education – interim report', Education Network for Environment and Development Tilbury, D. *Environmental Education Research* 1(2), 1995.

World Commission on Environment and Development (1987) *Our Common Future*, Oxford, Oxford University Press.

2 Local government, local people and Local Agenda 21

Jane Parker and Paul Selman[1]

Introduction

As environmental management emerged as a key issue during the 1970s and 1980s, so local government became one of the main agencies entrusted with its execution. Several researchers have identified the numerous initiatives of 'green planning' and environmental reporting which were introduced over this period (see Selman, 1996 for a summary). More recently, however, our concerns have extended from environmental management *per se* to the more complex issue of sustainable development, with its wider recognition of behavioural and institutional change, quality of life and community participation. Consequently, during the 1990s, we have witnessed a shift of emphasis from 'local government and the environment' to one of 'local governance and sustainability', implying a deepening and broadening of component issues.

Part of this change reflects our growing awareness of the inadequacy of top-down, command and control approaches to environmental management, and a recognition that it is social and political barriers, rather than scientific ones, which often prevent us solving environmental problems (e.g. Trudgill, 1990). Partly, it is also a recognition of the inseparability of traditional environmental issues from the wider social and economic concerns which affect our quality of life. An inescapable observation is that significant change towards sustainability will require more active citizenship, including individual life style changes, participation in creating and achieving community visions, and more accessible styles of local governance. As with all campaigns for enhanced citizenship, this is an elusive goal in an atomistic society, and one which is further compounded in relation to local environmental issues by the sheer insignificance of individual endeavour in the face of global dysfunction.

We will argue in this chapter that Local Agenda 21 is an especially promising initiative which can reconnect people with place, and help empower them in relation to key issues which affect their quality of life. It also relates to wider issues of citizen exclusion from effective participation in the scientific and technical debates of our 'risk society' (Beck, 1992), and thus to the quest for more inclusive and democratic local discourses (Irwin, 1995).

After a brief background to the research upon which this discussion is based, we will explore three connected points raised above, namely:

- the implications of sustainable development for the enhancement and extension of citizenship;
- the notion of civicness, in relation to the potential of Local Agenda 21 to engender individual and collective endeavour at the level of the community; and
- some initial results from empirical work arising from continuing research.

Background to the study

Our research is focused here on a small number of case studies of local authority sustainability programmes, and has a particular concern for their effectiveness in broadening the basis of citizen inclusion. LA21 programmes are often described as 'transparent, open and participatory' and, if this is anything more than fashionable rhetoric, we ought to find evidence of widening networks of community and individual involvement in sustainability initiatives. Our evidence is being gathered both by the analysis of documents and around 60 extended, taped, semi-structured interviews with key informants. Interviewees, identified mainly by a 'snowball' approach, have ranged from senior executives to environmental activists, and from academics to city councillors. Case studies from which primary evidence has been gathered on LA21s comprise Gloucestershire and Lancashire County Councils, Leicester City Council and Reading Borough Council.[2] Secondary analysis has also been undertaken of a project being conducted by the University of British Columbia (UBC) in collaboration with Richmond (BC) City Council, and of the 'Model Communities Program' being promulgated by the International Council for Local Environmental Initiatives.

The UK examples were selected for their various innovatory approaches: Gloucestershire for its early devolution of lead responsibility to a voluntary organisation, Lancashire for its long-standing record on environmental management, Leicester for its inclusion in the Environment City network, and Reading for its neighbourhood-based approach (supported by the Worldwide Fund for Nature UK). Our link to the UBC project gave us the opportunity to study the involvement of an academic department (the School of Community and Regional Planning) with a municipal council, and particularly their attempt to integrate their theoretical work on ecological footprints into planning policy and practice. Liaison with ICLEI further ensured that our deliberations have not taken place solely against a backcloth of advanced western municipalities.

Sustainable development and citizenship

As previously noted, there is a strong link between sustainability and active citizenship, as the necessary scale of social transition cannot be based purely on the delivery of services by central and local government. Consequently, a number of countries have instigated 'environmental citizenship' programmes, and the UK's national 'Going for Green' campaign reflects such an attempt to encourage greater commitment and responsibility from individuals.

Very briefly, there are two key underlying concepts relevant to the present discussion. First, it has long been recognised that citizenship may broadly be 'active' or 'passive', that is, requiring the individual to acknowledge responsibilities and duties which build up the general well-being of society, or enabling a person to enjoy certain political and other rights, respectively (see Hill, 1994, for a relevant review). Both of these are relevant to sustainable development: active citizenship expects involvement in environmental projects whilst, more passively, 'consumers' can benefit from initiatives such as the Citizens' Charter (which now incorporates a local government environmental *Citizens' Charter* as well as charters for environmental 'quangos'). Second, the history of citizenship has been one of an increasingly inclusive membership. Most recently, this has seen the inclusion of responsibilities towards the earth and nature as a component of 'good' (in contrast to 'deviant') citizenship. Indeed, many commentators concur that environmental considerations have recently given the ailing topic of citizenship a new lease of life (van Steenbergen, 1994). Another aspect is that the exercise of active citizenship, and the full enjoyment of citizens' rights, depend on individuals being adequately liberated from poverty and other forms of oppression. This, therefore, ensures that a wide range of quality of life and equal opportunities issues, and not only the traditional 'green' agenda, are being incorporated into sustainability programmes.

LA21 and civicness

It is evident from this brief review that the pursuit of sustainability cannot rely simply on service delivery or individual public-spiritedness – not least because the magnitude of problems requires us to make collective, and perhaps daunting, responses. The notion of the active citizen, as currently espoused by schemes such as 'Neighbourhood Watch' and 'Going for Green' is rooted in liberal-individualist ideology and seeks to combine a mix of self-help, voluntarism and philanthropy. Whilst recognising the contribution of voluntary activists, it is improbable that this provides anywhere near a sufficient solution. One way in which it will need to be extended is by engaging whole communities in sustainability initiatives, so that they take collective responsibility for their quality of life and the wider ecological footprint. This observation leads us to consider civic-republican and communitarian conceptions of citizenship – contested though they may be – as they emphasise that individuals become full citizens only by participating as members of a community.

Research at UBC has emphasised the importance of the social caring capacity (SCC) of a community, which reflects its stock of social capital in the form of community organisations and density of social networks. It could be argued that this is a characteristic British virtue, given the long tradition of its voluntary sector, and a strength to which sustainability programmes can play. Some research suggests that social capital is reflected as civicness and is fundamental to the successful implementation of policy initiatives – even more so than variables such as level of education, economic development and urbanisation. It is characterised by interdependent relationships, trust, reciprocity and co-operation, closely interwoven associations and social networks, and diversity among elected officials' backgrounds (Putnam *et al.*, 1993). Thus, it could be argued that the denser the social networks in a community, the higher the possibility of collective participation in sustainability projects. There are, of course, disputes within the communitarian case – imperfect correspondence between spatial and aspatial communities, weak coupling between levels of community awareness and actual civic activity, and variations in perceived attachment to a community amongst residents of a neighbourhood – but the general principles are likely to have some significant applicability (e.g. Malibeau *et al.*, 1989).

In relation to sustainable development, it has been cogently argued that past failures have been the product of perceived separation between cause and effect (notably the complex and obscure linkages between individual action and global environmental change) and of denial (refusal to contemplate the consequences or likelihood of environmental catastrophe). Effective LA21 programmes must both seek to reconnect individuals and communities with the consequences of their actions and open up channels through which they can participate in effective responses. These are ineluctably linked to civicness.

This is of more than theoretical interest. Planners and politicians have become increasingly concerned about the impact of NIMBYism on decision making, and are now generally aware that this is as much a product of citizen disenfranchisement as it is of selfishness. Consequently, there is a growing interest in innovative participatory techniques – such as visioning, consensus-building and round tables – which can yield mutually acceptable solutions to localised environmental conflicts. There is also a recognition of the value of networking best practice between environmental and community organisations, both because of the relatively trusted nature of information when it is communicated informally between equal partners, and because this is an efficient means of sharing limited resources (Martin, 1995).

If the benefits of these approaches are to be realised, it is axiomatic that we must move up the 'ladder of citizen participation' (Arnstein, 1969), from traditional consultative plan making and informative State of Environment reporting, to rungs where citizens retain genuine influence. Whether their authors recognise it or not, this is what LA21s are generally seeking to achieve. If they are succeeding in this mission, they should be introducing denser vertices between organisations, widening networks of knowledge and influence, and bringing forth new community champions. Otherwise, they may simply be attracting the traditional meetings-junkies, burdening existing volunteers with additional responsibilities and 'rebadging' current environmental projects. In essence, the latter situation may not be too serious, as there may be an inherent upper limit to the level of social capital within a locality, whilst pre-existing initiatives are likely to have a continuing validity. Nevertheless, the effectiveness of LA21 in diverting both processes and products towards a more radical purpose is of fundamental significance.

Some key issues

At the time of writing, our empirical work is incomplete and we have not had the opportunity for leisurely reflection on its implications and significance. However, some key issues surface as recurrent themes. These include:

- the broadening of the environmental agenda;
- incorporating locally derived priorities;
- creation of equal partnerships for implementation and financing;
- achieving a consensus;
- civicness and the committed individual; and
- barriers to environmental action.

Broadening the environmental agenda

It is increasingly clear that LA21 is proving to be, both actually and potentially, about much more than environmental management. In particular, it is involving initiatives which move communities up the 'ladder of citizen participation', strengthen aspects of local governance, introduce denser vertices between organisations, and widen networks of knowledge and influence. This is largely taking place as an evolutionary process, and the pace with which it is occurring is variable. It has been common practice for local authorities to commence the LA21 process by putting their own houses in order, and build on the environmental management practices established during the 1980s; most have subsequently recognised that sustainable development encompasses a spectrum of critical quality of life issues. There is an implicit recognition that the LA21 process has the capacity to yield an alternative citizens' agenda, alongside the more normative local authority agenda.

This broadening agenda has been handled differently by individual case study areas. It is interesting to compare the approaches of Lancashire County Council and Reading Borough Council. Lancashire has made a concerted effort to capitalise on the use of its well-resourced method of 'state of environment reporting' initiated by its Environment Unit (established well before the 'Rio' summit). As such, the County Council has not felt that LA21 has had a mould-breaking impact per se, and there is a concern not to jettison the environmental work which had already occurred. However, the Environment Unit has received some criticism for not addressing the participative element sufficiently early, and this may have affected the overall dynamic of the process, though it is now starting to be rectified. In Reading, the Council and the World Wide Fund for Nature (WWF) UK entered into an experimental partnership and created, at an early stage, a specific LA21 team; strong collaborative links were established with the community education arm of WWF UK from the outset. Thus, sustainability work was differentiated to a large extent from previous environmental management practices, and the new team has gradually become much more than a re-badging of its predecessor Environmental Strategy team. The ability of the new team to incorporate experts on an inter-departmental basis (e.g. from environmental services, education, recycling, community development) has been facilitated by the authority's overall corporate style of working.

Reading's greater potential to reflect, at an early stage, an emergent set of community-based agendas is also related to the scale of the initiatives. The County Council in Lancashire has, naturally, maintained a county-wide focus to its work, though most of the District Councils are also preparing their own independent LA21s, and the county now includes two pilot areas for a 'Going for Green' community participation programme. Reading's approach has been characterised by a strong neighbourhood emphasis since its inception, with GLOBE ('Go Local On a Better Environment') groups being established in specific wards of the borough (Howells *et al.*, 1995). The scale of operation has obvious repercussions for the success of initiatives seeking to enhance levels of social capital within communities, and this aspect will be reviewed subsequently.

Success in deriving a citizens' agenda can also be gauged by the extent to which initiatives are incorporating locally derived priorities, drawing on networks of expertise, being implemented and financed between equal partners, and being legitimated on the basis of a broad consensus. All of these elements recur regularly.

Incorporating locally derived priorities

The incorporation of locally derived priorities and use of networks of expertise are evident in Lancashire's 'Environmental Action Plan' and Environmental Information System, Reading's various GLOBE group Action Plans, the Blueprint for Leicester, and Vision 21's *Sustainable Gloucestershire – the Biggest Issue.* Within ICLEI's 'Model Communities Program', also, the participating municipalities are all producing comparable statements (ICLEI, 1996). All these documents vary greatly in the extent to which the local priorities represent the views of laity or experts, and this in turn has affected the level of ownership felt by the community towards these agendas and their positive use as citizen agendas. Several informants anticipate that the most durable ones will be those which are able to integrate lay and expert input, though the difficulties of genuinely collaborative working between these two constituencies are widely acknowledged. Participants in Gloucestershire's 'Vision 21' process remarked that the process has gone some way towards this integration, with an emphasis on 'expertise and not experts'. All participants are encouraged to take part in the process as individuals and not as representatives and, in this way, it is hoped that networks will more readily form across traditional boundaries, and that over-commitment, followed by disillusionment, can be avoided. By contrast, some respondents expressed a degree of disappointment over the first Lancashire 'Environmental Action Plan', in which a network of county-wide experts (members of a large Environment Forum) were, without having a real sense of ownership, encouraged to sign up to pledges, many of which were in reality difficult to keep. A complementary network in Lancashire has been the Centres of Environmental Excellence, which avails differing aspects of expertise and good practice to a variety of audiences (e.g. education and industry). Leicester's 'Blueprint' process has, to date, sought to draw in expertise, particularly through various existing networks in and around the city – concerned, for example, with community safety or elderly people – and this has also incorporated both lay and expert participants. This process was favourably commented upon by the local media, although local political changes created something of a hiatus which caused temporary though significant problems of continuity. Whilst the extant networks have continued their specialist work, engaging individuals from these separate networks to work together on broader visions for the city has been problematic, particularly where this has involved experts and non-experts trying to conduct a representative and broad issue-based dialogue.

Partnerships for implementation and financing

There are now examples of implementation and financing between equal partners taking place at a variety of levels. In Leicester, the City Council and Environ (an independent environmental charity and consultancy) share responsibility for the co-ordination of the 'Blueprint' process. At a different level in Leicester's process, a new partnership (Leicester City Partners) includes leading city figures, and comprises leading councillors, the bishop, university vice-chancellors, president of the chamber of commerce and voluntary sector leaders. This group is contributing to 'Blueprint' by devising collaborative strategic visions for the city, which will then be meshed with the community-based visioning process, prior to implementation.

Achieving a consensus

Different case studies have employed different means of attempting to achieve a consensus over complex sustainability goals, and this has been associated with various degrees of perceived success and failure. For instance, the initial neighbourhood-based phase in Reading is starting to be welded into a town-wide layer, and all GLOBE groups have expressed a keen interest in working with their equivalents elsewhere in the Borough. The Council's LA21 team have selected the divisive issue of traffic/transport as an initial focus for collaborative engagement. It is hoped that the use of various community development techniques can facilitate the attainment of consensus over priorities on this issue, and avoid descending into NIMBYism. This ongoing project has brought together a range of community and Council representatives, who in turn have identified the need to include private-sector operators. Gloucestershire is employing innovatory techniques such as future-search in an attempt to facilitate consensus and develop shared visions at its stakeholder conferences (on housing, economy and transport). One participant in the Accommodating Gloucestershire Conference (confronting the county's highly contentious issue of new housing provision) noted that 'we have established much common ground', whilst the Conference Report affirmed the success of 'bringing people together in a format where they could find ways to harness their enthusiasm and commitment towards solving the problem of accommodating Gloucestershire in the years to come' (Vision 21, 1996). From these early foundations, several self-generated projects involving Vision 21 and conference delegates have started to emerge.

Civicness and the committed individual

The more inclusive methods used to derive local agendas doubtless concur with much popular rhetoric about ownership and empowerment. However, it is essential that participatory processes reinforce, rather than undermine, the legitimacy and authority of local government. A positive venture, therefore, is that of Reading Borough Council in providing some of its officers with special group facilitation training by the Community Education Development Centre. These officers can then assist the production of neighbourhood action plans, by facilitating local residents to channel their aspirations for their locality. Since these often come in the form of demands or complaints targeted at the local authority, officers are trained in dealing with this, so as not to raise expectations amongst the community and to avoid becoming overburdened themselves with promises that cannot be kept. This communicative relationship based on informed trust has achieved notable early success in most pilot areas, with both officers and community members identifying positive personal rewards from the process, as well as tangible outcomes for the locality. At a more strategic level, the process has strong political commitment from the leadership and is seen and publicised as a first step in strengthening the legitimacy of the Council amongst the citizens of the borough. However, Leicester's ambitious consultative process which led to the production of its 'Blueprint' received some unfavourable comments arising from its failure to deliver on raised expectations. These shortcomings are generally attributed to political hiatus and a consequent change to the resourcing of the LA21 initiative, and thus perhaps attests the fragility of such processes within the wider political context.

Experience strongly suggests that citizens' agendas will not transpire simply by council officials and elected members 'letting go', but by the assiduous application of a specific range of challenging techniques. We have noted the use of techniques such as facilitation training, consensus building and visioning. It is interesting, though not surprising, that the use of such techniques is taken for granted in Gloucestershire's 'Vision 21': indeed, the process seems to have found its natural home in a largely pre-existing network of individuals with a wide skills base in such techniques. This circle is generally outside the local authority domain, and the skills are perceived as new ones to be acquired by council staff, and are greeted (not surprisingly) with some scepticism by elected members. In the case of 'Vision 21', this emphasis on new techniques may be hampering its influence in some areas where there are more conventional modes of working, although there are examples of previously uninitiated individuals finding new techniques refreshing and rewarding.

The evidence, therefore, strongly confirms that the role of local government in LA21 processes cannot simply be defined in terms of the relative desirability of top-down or bottom-up processes. One of the UBC researchers referred to the inevitable duality between the need for goal setting and central planning on the one hand, and for local implementation and the incorporation of indigenous preferences on the other. Furthermore, part of the enthusiasm for LA21 has been its emphasis on participatory models of democracy, and some of its most committed activists have seen it as a means of displacing the rather tired model of representative democracy within the local state. However, local government is clearly a major stakeholder in sustainability programmes – for example as a financial broker, source of champions, a service provider and custodian of civic tradition – and it would be counter-productive for it to be marginalised. What is required is a splicing of the models of representative and participatory democracy, and of the mechanisms of popular agendas and official plans.

Successful splicing of representative and participatory modes of democracy will require that LA21 has a strong community, even neighbourhood, focus. Whilst it is acknowledged that the term community must be used with caution and the concept need not necessarily refer to a spatial community, this does not detract from the fact that participatory techniques, whether applied in spatial communities (neighbourhoods) or within communities of interest, can rarely produce a fully representative agenda, and are thus an insufficient alternative to representative democratic processes (Young, 1995). Their greater value, however, lies in enabling LA21 practitioners to identify sources of social capital. For example, 'Vision 21' is, at the time of writing, embarking on a sectorally based outreach programme, utilising visioning methods to identify stocks of social capital in the form of community organisations and density of social networks that may assist in pursuit of sustainability. Outreach work is thus being targeted on a county-wide basis to women's groups, the unemployed, older people, the media, and so forth. By contrast, Reading's neighbourhood-based GLOBE groups have sought to capitalise on social capital as it occurs within networks of people located in specific neighbourhoods. We have considered the channels through which the GLOBE message has been permeating the community beyond the 15 to 20 individuals who regularly attend meetings in each area, and it would appear that membership of other community groups and wider participation in practical tasks on action days are key factors in securing and sustaining involvement. It will be especially interesting to see how the neighbourhood-based projects can sustain themselves in the medium to long term, as council officials step away from facilitating individual processes whilst still providing a strategic lead, and hand over group facilitation and resourcing to members of the community. Early observations imply that some groups will be more successful than others, and it will be interesting to explore the elements of social capital which underpin this relative performance.

Two findings of the UBC researchers warn of the dangers of excessive reliance on community participation, especially in the absence of adequate structural mechanisms for support. They note, for example, the potential for initiatives to produce perfunctory participation, characterised by overworked volunteers and staff repeatedly taking on additional roles to the detriment of actual programme implementation. Furthermore, the problems of finding volunteers for the labour-intensive work of attending working parties or organising neighbourhood activities typically led to an over-representation of more affluent and retired people and professionals. Similarly, there can be tensions and accusations of NIMBYism when the central funding authority and the local population have different priorities. Despite this, it is inevitable that some individual citizens will need to donate remarkable quantities of personal energy, time and commitment to the process, and these people will often be crucial to the success or failure of an initiative. They can be critical to harnessing extant or generating new social capital, be instrumental in providing structures to support the LA21 process, or be key political figures, providing initiatives with legitimacy and endorsing a representative model for the promotion of civicness and LA21.

Key individuals are often referred to as champions, and we have observed that champions can emerge from various sources, including local neighbourhoods, NGOs and even major bureaucracies. Some elected members clearly perceive themselves as the real community champions. What appears to matter is that a competent and committed individual is willing to carry through a project, be it the whole LA21 process or just one aspect of it, from initiation to completion. One respondent was quite clear about the magnitude and duration of a personal commitment, and stated that champions must be prepared to 'walk the talk'.

Similarly, some individuals are identified as catalytic personalities who become agents of motivation and change. We have noted, for example, the influence of particular leaders of the Council in both Lancashire and Leicester, although it is typical to refer to catalysts as arising more spontaneously from non-governmental quarters. However, we should bear in mind that, in chemistry, catalysts speed up, but remain unaffected by, a reaction, whereas social catalysts may 'burn out' under the burden of expectation. Further, when such a person departs temporarily or permanently a project may fail 'if there is no structural mechanism for the catalysis to continue' (to quote one Richmond interviewee). A clear message is that LA21 – or, indeed, any programme dependent on civicness – must incorporate support mechanisms to release individuals from unreasonable expectations and to facilitate continuity.

In at least one of our case studies there was evidence of a further key agent – the well-connected super-networker. This is perhaps a hallmark of sustainability, as the overall concept attracts, and is perhaps comprehensible to, only a minority of people, whereas individual components of sustainability programmes draw upon their local networks of single-issue participants. The super-networker will have the imagination, interest, time and energy to take an active role in co-ordinating groups and in liaising with several parallel streams of interest. Not only do they underpin strategy synthesis but they also typically bring their personal networking skills to bear on the mobilisation of financial and personnel resources. We have encountered super-networkers in various guises and at various levels in all of our case studies, and it is interesting to note that they often possess a background in areas such as psychology or voluntary work, rather than environmental science.

Barriers to environmental action

A further theme to emerge from our fieldwork has been the recognition of impediments to effective environmental action, even where there is general acknowledgement that municipal action is desirable. During the early growth of concern for environmental management, these problems were widely assumed to be scientific, such as lack of information or of technical solutions. However, it is now accepted that the most intractable impediments are often social and political ones, and we have attempted to demonstrate the ways in which sustainability programmes are trying to target ways of overcoming these.

Researchers at the University of British Columbia considered the barriers to fall into three categories, namely, perceptual/behavioural, institutional/structural and economic/financial. More specifically, these reflected weaknesses of: comprehension of the issues, legal and administrative powers, funding, prioritisation over competing issues, and constituent support. Other respondents have alluded to lack of 'buy-in' from key departments, inherent difficulties associated with multi- and inter-disciplinary working, the perception of participatory mechanisms as threatening normal practice and the problems associated with securing a sense of shared ownership of the process.

More encouragingly, interviewees have identified a number of gateways as well as barriers to implementation. These appear to include:

- presence of a locally receptive culture (often associated with a tradition of voluntarism and participation);
- focusing on a limited set of proposals and recommendations (though this may conflict with locally derived agendas);
- close working between civic staff and community groups;
- improving networks;
- having clear objectives, timelines and resource expectations;
- including structural mechanisms which support active participants; and
- political commitment to an opening up of council processes.

Conclusion

It is clear that it is insufficient for LA21 to be just another document, setting out a local authority's environmental proposals for its administrative area. It is a process, perhaps even a threatening one, which must involve localities in real change. The capacity for and nature of this change will vary according to the characteristics of local place – its environmental character, comparative importance of local quality of life issues, and socio-economic and ethnic composition, for instance. The basis for change will inescapably be that of partnership, in which all parties or stakeholders become vulnerable to the possibility of personal challenge. The key partners will be drawn from local government, official agencies, NGOs, community-based organisations and individual citizens, and the process will involve splicing together the authority and representativeness of local government with the enthusiasm and knowledge of neighbourhood participants. Much wider use will start to be made of techniques for visioning, consensus building, capacity building and participatory environmental assessment, yet these may only be the tip of the iceberg in relation to more pervasive discourses on democratic reform and the risk society.

There is a rejoinder which we should not ignore, however. Whilst much of the content of LA21 conforms to currently popular rhetoric on grass-roots, inclusive, indigenous processes, there is a real possibility that apparent empowerment may prove to be hollow in practice. Civicness may be exploited merely as a cheap option, consensus may prove impossible and cynicism may rapidly replace enthusiasm when implementation fails. Real results may involve spending considerable sums of money and passing effective legislation, rather than relying on the 'good environmental citizen' willingly to make personal sacrifices and life style changes. This is not to negate the pivotal importance of civicness, but it does remind us that motivation, by itself, is insufficient.

Notes

1 This chapter arises from research conducted under ESRC grant No. L320253221, 'Policy, process and product in Local Agenda 21' (a project within ESRC's Global Environmental Change Programme, Phase 4).

2 The key Local Agenda 21 publications for the case study areas are: Gloucestershire County Council/ Vision 21, Sustainable Gloucestershire – the Biggest Issue; the Vision 21 Handbook for Creating a Brighter Future (1996); Lancashire County Council, *Lancashire Environmental Action Programme – a Local Agenda 21 for Lancashire*, (1993); Lancashire County Council, *Lancashire's Green Audit 2: a Sustainability Report* (1997); Leicester City Council, *Blueprint for Leicester – Findings Report* (1995); Reading Borough Council, *Reading's Agenda 21 Statement – What's it All About?* (1997).

References

Arnstein, S. (1969) 'A ladder of citizen participation', *Journal of the American Institute of Planners*, 35, 216–24.

Beck, U. (1992) *Risk Society: Towards a New Modernity*, London: Sage.

Hill, D. (1994) *Citizens and Cities: Urban Policy in the 1990s,* Hemel Hempstead: Harvester Wheatsheaf.

Howells, C., Hollins, M. and Percy, S. (1995) 'Reading's Neighbourhood Agenda 21 a unique approach to sustainable development?' *Local Government Policy Making*, 22 (2): 48–52.

International Council for Local Environmental Initiatives (1996) *The Local Agenda 21 Planning Guide: an Introduction to sustainable development Planning*, Toronto: ICLEI.

Irwin, A. (1995) *Citizen Science: a Study of People, Expertise and sustainable development*, London: Routledge.

Malibeau, A., Moyser, G., Parry, G. and Quantin, P. (1989) *Local Politics and Participation in Britain and France*, Cambridge: Cambridge University Press.

Martin, S. (1995) 'Partnerships for local environmental action: observations on the first two years of Rural Action for the Environment', *Journal of Environmental Planning and Management*, 38: 149–65.

Putnam, R., Leonardi, R. and Nanetti, R.Y. (1993) *Making Democracy Work: Civic Traditions in Modern Italy*, New Jersey: Princeton University Press.

Selman, P. (1996) *Local Sustainability: Managing and Planning Ecologically Sound Places,* London: Paul Chapman.

Trudgill, S. (1990) *Barriers to a Better Environment: What Stops Us Solving Environmental Problems?*, London: Belhaven.

3 Getting people involved

Walter Leal Filho

Introduction

On an international scale, many governments, at national, regional and local levels, have found that, as far as the environment is concerned, cure is far more laborious and expensive than prevention (Bernstam, 1991; Brat and Steetskamp, 1991). Aware of that fact and recognising that public participation in environmental conservation is critical to health maintenance (WHO, 1995), sustainable development (Templet, 1995) and economic growth (OECD, 1993; Schmidheiny, 1992), many governments, national and local level included, have been trying to solve their environmental problems by trying to create a broader awareness about them, mostly through one-off information programmes. But with little consideration to the need methodically to inform the public, raise their awareness on a certain issue, inform on their variables and then seek changes in attitudes or behaviours, few such information programmes have been or are likely to succeed.

Part of the problem is due to the fact that there are seldom attempts to co-ordinate environmental information and education programmes, both at the formal and non-formal level, and to integrate them with initiatives aimed at promoting the conservation of the environment at the level of municipalities and communities. Taking into account the reality in the majority of the industrialised nations in Europe, where environmental education initiatives are found at a rather advanced stage of evolution (Leal Filho, 1996a), it is clear that good opportunities to capitalise on existing resources to the advantage of the conservation of the local environment are being missed. Some of the reasons which explain such a state of affairs are:

- environmental conservation programmes have traditionally given little or no emphasis to the need to incorporate an environmental education dimension;
- education as a whole and environmental education in particular, when looked at critically, are in the rhetoric of politicians but are nonetheless hardly ever prominently seen as parts of action plans;
- the perception of the wider public, and especially among decision makers, is that environmental education and awareness initiatives are limited to formal

teaching and to primary or secondary schools, therefore lack a practical application to other sectors of society.

The latter view is of course a misrepresentation of what environmental awareness and environmental education should and indeed can do to help implement environmental conservation policies and to foster public participation.

Since the publication of the Brundtland Report (WCED, 1987), the subsequent UN Conference on Environment and Development (UNCED) held in Rio de Janeiro in 1992 and the endorsement of Agenda 21 by over 100 Heads of Government, the call for environmentally sound or sustainable development has been echoed in industrialised and developing countries the world over (UN, 1992) and the search for indicators of sustainability has intensified. The past three years or so have seen an unprecedented opening of opportunities for the implementation of environmental conservation programmes aimed at the ultimate goal of sustainable development, despite the fact that there are still some areas in Africa, Asia and Latin America where conservation of the environment and strict environmental policies are usually seen as obstacles for economic growth.

This volatility has a powerful impact on the ways countries pursue environmental policies. When economic output and levels of employment are high, the introduction of restrictive environmental policies becomes a difficult task. Although not unheard of in industrialised countries, such a situation is a more serious problem in emerging countries where less rigid environmental policies and legislation allow industrial activities not likely to be tolerated elsewhere. It takes a brave politician in a poor nation to say no to the setting-up of new industrial activities or enterprises on the grounds of environmental problems or environmental risks.

However, modern environmental conservation is no longer a matter for governments alone. The public, as documented elsewhere (see Cracknell, 1993; Dunlap and Van Liere, 1984; Pirages and Elrich, 1974), is willing to become more and more involved and this is good news. Taking into account that a major challenge faced by local authorities in their attempts to implement the recommendations of Agenda 21 as a whole, and Chapter 28 in particular, is how to get people to play along and be active in efforts towards the conservation of the local environment in both European countries such as the UK (Local Government Management Board, 1995) and elsewhere, it can be said that the systematic implementation of environmental education can be one of the ways of achieving that goal. To the same measure, it can be seen that without public interest and participation, even the best designed environmental conservation programme is not likely to succeed, hence there is a need to identify approaches and methods through which public interest and participation in environmental conservation efforts are encouraged.

Patterns of participation in environmental affairs

On a world wide basis, public participation in environmental conservation, although regarded as important, has not always been effectively pursued. An example comes from the international efforts towards the rational use of water. As stated by Landrigan (1997), in December 1995, in an effort to rationalise the scattered array of water programmes, 75 representatives from 56 agencies and governments around the world gathered in Stockholm at the First Water Meeting. The Meeting, which was hosted by the Swedish International Development Agency (SIDA), was organised by the World Bank and the UN Development Programme (UNDP), which have worked extensively on water projects.

The involved agencies felt that progress in translating into practice the principles articulated at the Dublin International Conference on Water and the Environment in 1991, which were subsequently reiterated in Rio, at Habitat II and other fora, was disappointingly slow. In addition to the usual problems in implementing international agreements, such as limited budgets, lack of political priority and difficulties in setting-up new infrastructures, the real value of water conservation efforts was not fully appreciated in some countries; for example, whilst there may be an interest in water conservation, it may not be strong enough to be matched by the investment needed to address the problem. There is, in other words, not enough motivation to put ideas into practice and to involve sectors such as local administrations, industry and women in water conservation initiatives. Yet, without the involvement of these groups and others such as farmers, little progress can be expected. This attitude can also be seen in relation to other environmental agreements, and is symptomatic of the need to raise support, from the top to bottom as well as from the bottom to the top, for the conservation of environmental resources, and to try to ensure that environmental information is duly gathered and disseminated (Kremers and Kraseman, 1996).

Similar to what happens with decision makers (at the top), if people (at the bottom) have a sense of hopelessness whereby they feel that whatever they do will not change anything, then conservation efforts, no matter how well intentioned, are likely to fail. Instead of being positive, government efforts will in fact be pernicious. Many people will simply view a proposed initiative as another attempt to interfere with their lives and as a threat to their quality of life and access to facilities. Three examples of conservation projects in Indonesia, Peru and Costa Rica, where ideas discussed in government buildings without public participation were designed and implemented, met with substantial local resistance. In some more extreme cases, such as the Narmada Dam in India, a planned development scheme could not be implemented at all, such was the degree of public frustration.

The extent of public participation in environmental affairs varies from country to country and in some countries the support of the media plays a key role (Anderson, 1991), although this contribution should not be overestimated (Gaber, 1993). Even in organised groups of nations, such as the EU, the level of interest on environmental matters as a whole and on local environmental initiatives in particular, differs significantly. In some countries, such as the UK, Denmark, Finland and Germany, there is generally a strong interest in and concern about environmental affairs, translated by an emphasis on the handling of environmental themes in schools (in addition to non-formal education), while in countries such as Greece or Portugal, environmental matters are perceived as relevant but environmental awareness in relation to the problems caused by tourism was undertaken in Portugal (Leal Filho, 1994a). In the developing world, Thioune (1993) tried to promote awareness on the use of non-formal educational approaches to raise the profile of environmental conservation in rural areas in Senegal. In Europe, local environmental initiatives have been supported by awareness-raising programmes such as the Lancashire Environmental Action Programme (Lancashire Environment Forum, 1993), which has a chapter devoted to education and awareness, and the Lahti Environmental Forum in Finland (Lahti Environmental Forum, 1996) which sees enhanced environmental awareness as a vital component for the successful implementation of environmental policies.

Similarly, there are plenty of examples of initiatives aimed at promoting environmental education, some which have been outlined in publications such as *Trends in Environmental Education Worldwide* (Leal Filho and Hale, 1994), *Environmental Education in Small Island Developing States* (Leal Filho, 1994b) or *Environmental Issues in Education* (Harrisson and Blackwell, 1996), including publications produced in languages other than English, such as Schleicher's (1996) *Environmental Awareness and Education in the European Union* (*Umweltbewußtsein und Umweltbildung in der Europäischer Union*). The history of environmental education is also punctuated by meetings such as the Belgrade Workshop (1975), the T'bilisi and Moscow Conferences (1977 and 1987), not to mention the Earth Summit itself, in 1992, in which, with no exception, the need for introducing environmental education at the local level was emphasised over and over again.

Reasons for participation in environmental conservation

Listed below (in no particular order) are some of the most common reasons for public participation in environmental conservation efforts.

- the matter is perceived as relevant by individuals;
- individual participation is seen as making a difference;
- motivation is provided by authorities, family, friends or peers;
- financial benefits;
- social benefits;
- improvements in infrastructure;
- direct profit;
- concern for the environment.

Unfortunately, although the penultimate reason, namely direct profit, may ultimately direct a number of people to take part in conservation efforts, the final reason, namely concern for the environment, is not often seen as the main reason why individuals get involved in environmental conservation programmes. This is partly because, concern per se, is in nearly all cases not enough to motivate action, other elements are needed (e.g. the direct impact of an environmental problem on one's health) to influence an individual's decision to act. The role of motivation, as described earlier, should also be taken into account. It is one of the challenges for environmental education and awareness initiatives, to change this pattern.

There are many elements that might influence an individual's decision to take a particular decision in relation to the environment. In choosing which product to buy, which way to go to work, the type of transport to use or even in deciding which type of lamp bulb to use at home, all have a connection with the environment, which is sometimes, due to the weight of financial or political pressures, ignored. For example, the decision to buy an environmentally friendly (i.e. energy saving) bulb is easier to make if it does not cost much more than a normal one. The same applies to lead-free petrol and hairsprays.

From the many issues that may influence an individual's decision to become involved in environmental affairs, five main items (taken from international examples of successful local environmental programmes) appear to be very closely associated with positive outcomes. It would seem that such variables, when duly considered, may also be relevant at the local level and may provide some assistance for local authorities looking for ideas on how to make their local environmental conservation plans work.

Provision of information

This is the basis for public involvement in environmental issues, for without information there can be no expectation that a particular initiative may be successful. People have to know the what, why and for what of environmental conservation initiatives, and only on the basis of proper answers can some degree of interest be expected. The provision of information, although sometimes implying costs for a local authority, is infinitely cheaper than having to deal with the public's resistance to a proposed project later on. The social and environmental impacts of the Narmada Dam in India, for example, were not clearly outlined to the local population, which saw the dam as a threat to their livelihoods and to their own existence. Perhaps such resistance would be reduced if people were properly consulted and informed from day one.

Motivation

This is also an important component, for the provision of information has only an initial effect since without motivation actually to change action or behaviour, efforts are likely to fail. The same also applies to the traditional command and control approaches, whereby instructions are given and people are expected to follow them. A typical example of lack of motivation to undertake environmental conservation initiatives was in England in the summer of 1995, when, despite being instructed by the water authorities not to water their gardens, most citizens chose to ignore this, since they were not motivated enough to make their own, albeit small, contribution to water-saving efforts.

Commitment

Commitment here is interpreted as the interest in pursuing an initiated change of behaviour or action, in the long term. Very often changes of behaviour occur in the short term and people soon tend to go back to their old habits. For example the decision taken in the mid-1980s by the Athens authorities to curb air pollution by restricting car traffic in the city according to the licence plates (odd numbers on certain days and even numbers on others), was duly followed in the first six to eight months. After that, people started to change their licence plates (illegally) or use a second car, which was usually much older and therefore much less fuel efficient.

Incentives

Incentives towards environmental conservation may be derived from various instruments that generally have environmental benefits. In many countries, environmentally sustainable products and technologies are seen to be a prime factor in mobilising public support for them (e.g. the use of low-temperature and low-radiation lamp bulbs). A good example is the *pfand* or deposit system in Germany, whereby people buying drinks in glass bottles pay a deposit, whose value is included in the product's price. This is aimed at encouraging recycling, and millions of German households actively take part in such a simple but effective way of conserving natural resources. The reason why such an approach is not more widely practised is the lack of interest from industry and/or from decision makers in introducing similar schemes.

Some social and environmental specialists world wide are following developments related to the analyses of these variables very closely, while others have been relating these to other major changes in public perceptions (see Habermas, 1989). Part of the reason for this is the fact that the vast majority of policies listed in National Environmental Action Plans (NEAPs), Environmental Assessments (EAs) or in Local Environmental Action Programmes or Plans (LEAPs) require consultations with NGOs and with the public. At the same time, practitioners and local authorities alike recognise that citizen involvement at all levels, and their active participation, is fundamental to all forms of effective environmental protection, including those related to areas such as urban pollution control and waste management, which lie at the very heart of local environmental conservation initiatives.

However, the emphasis on information, motivation, commitment and incentives does not solve another problem that afflicts local authorities and slows down attempts to enhance the effectiveness of local environmental programmes, namely poor co-ordination (Leal Filho, 1997). Unless co-ordination between the various offices whose work is related to the environment (e.g. the environment division, the roads division, the infra-structure division and so on) is practised, efforts are bound to be duplicated at best and wasted at worse. A certain degree of co-ordination between the various parties is needed, to enable individual initiatives to exert a maximum impact and reduce the risks of duplication or hitting wrong targets.

Overcoming financial barriers

A common issue in addressing the problems outlined here, especially in the less economically developed countries of central and eastern Europe (e.g. the Czech Republic, Poland, Hungary), has been the lack of funds for pilot projects and capital funding. Due to the significant implications this has on making ideas turn into reality, it deserves special emphasis.

Aware of the problems in the field of financing and the impact this can have on a country's capacity to move forward, a recent call for projects was made by the Commission of the European Communities – the executive branch of the European Union – making significant amounts of funding available for local initiatives which focus on the environment. The initiative provides a mechanism for financing innovative regional development measures in collaboration with countries and regions outside the EU. Entitled Ecos-Ouverture, the programme, launched in 1991, has invested 48 million ecus from the European Regional Development Fund (ERDF) and from Phare programme resources (a European programme which funds projects bridging East and West Europe). During the programme period starting in 1997/98 the Ecos-Ouverture programme will receive 17 million ecus from the ERDF and 7 million ECUS from the Phare programme, enabling it to make financial contributions towards mutually beneficial co-operation initiatives between regions and towns in the EU and their counterparts in central Europe, the newly independent states (NIS/FSU), and the non-member countries of the Mediterranean (Algeria, Cyprus, Egypt, Israel, Jordan, Lebanon, Malta, Morocco, the Palestinian territories, Syria, Tunisia and Turkey). The following five fields have been selected for the EU's latest call for proposals:

1. improving the working methods of local or regional authorities through the establishment of local or regional development strategies and urban and regional services;
2. improving access to the European market for small businesses in the areas concerned, particularly through appropriate techniques for co-operation between firms and improving the supply of services to small businesses which encourage them to innovate while developing the role of local administrations;
3. developing specific local potential, particularly for the creation of permanent jobs;
4. establishing and developing resource centres to promote equal opportunities in economic life;
5. preserving and improving the environment with a view to sustainable development, by promoting renewable sources of energy and energy saving.

Co-operation projects should cover between three and seven different areas in at least two Member States and at least one partner country. The expected results include progress on the ground which provides a vehicle for the exchange of experience or transfer of know-how, and which can also have a demonstration effect for the region and neighbouring regions. In addition, it foresees the development of a culture of co-operation at regional and/or local level and an increase in the capacity for action of local organisations and persons engaged in economic development. Moreover, the development of work plans and feasibility studies for measures to be financed subsequently and the preparation of applications for funds from appropriate bodies which can provide finance, whether within the Community or elsewhere, are also envisaged.

The fact that the environment is one of the priority areas means that such an opportunity, as an example of what is and what shall be available, should be used so as to break down the inertia seen in relation to the actual implementation of local environmental conservation and local environmental education programmes. Similar lines of financing are available for EU countries themselves and for co-operation between European countries and countries in Africa, the Caribbean and the Pacific (the so-called ACP countries), as well as for joint projects with Asia (e.g. INCA (International Co-operation with Developing Countries)) and Latin America (e.g. ALPHA (a programme to support scientific research with partners in the EU and developing countries)). They illustrate that even problems with funding can be overcome and should not pose an insurmountable barrier in catalysing people's involvement in local environmental initiatives.

Conclusions

Most local environmental initiatives are committed to promoting a variety of innovative ways of fostering environmental conservation that yield global environmental benefits. Some environmental awareness and education programmes have been introduced sporadically in the past few years, in part because initiatives require more attention and resources than some local authorities are prepared to invest. Lack of funding, especially in poorer countries, has been mitigated over the past few years through special funding initiatives aimed at supporting local environmental work through co-operation between EU and non-EU countries.

Much of what has been described in this chapter will have implications for policy development and bear some connection to the work performed by local, as well as national and international organisations. Since these comments are made on the basis of practical research, having therefore a close link with reality, it is hoped that some ideas might be incorporated as part of the work of such organisations *vis à vis* enhancing their capability to gather further support.

To pursue the promotion of public involvement in local environmental programmes, the challenge is not so much how to find the money (which still remains a basic issue), but how to foster that interest and monitor its development, ensuring that the interest does not wane and at the same time learning from successes and failures. Paying due attention to the four variables outlined in this chapter, and the need to look at ways of generating financial support, may be one way of proceeding, together with consolidating best practices, understanding gaps that need to be filled and developing more strategic implementation plans for the future.

References

Anderson, A. (1991) 'Source strategies and the communication of environmental affairs', *Media, Culture and Society* 13 (34): 459–76.

Bernstam, M. (1991) *The Wealth of Nations and the Environment*, London: Institute of Economic Affairs.

Brat, L. C. and Steetskamp, I. (1991) 'Ecological economic analysis for regional sustainable development', in Costanza, R. (ed.) *Ecological Economics: the Science and Management of Sustainability*, New York: Columbia University Press.

Cracknell, J. (1993) 'Issue arenas, pressure groups and environmental agendas', in Hansen, A. (ed.) *The Mass Media and Environmental Issues*, Leicester: Leicester University Press.

Dunlap, R. E. and Van Liere, K. (1984) 'Commitment to the dominant social paradigm and concern for environmental quality', *Social Science Quarterly* 65: 1013–28.

Gaber, I. (1993) 'A cold shoulder for the environment', *British Journalism Review* 4: 4.

Habermas, J. (1989) *The Structural Transformation of the Public Sphere*, Cambridge: Polity Press.

Harrisson, G. and Blackwell, C. (eds) (1996) *Environmental Issues in Education*, Aldegate: Arena.

Kremers, H. and Kraseman, H. (1996) *Umweltdaten verstehen durch Metainformation*, Marburg: Metropolis Verlag.

Lahti Environmental Forum (1996) 'Lahti City Environmental Management Scheme', *Lahti Today* 4, Sept.: 1.

Lancashire Environment Forum (1993) *Lancashire Environmental Action Plan*, Lancaster: LEF.

Landrigan, S. (1997) 'Solving the water crisis together', *Environment Matters* Winter/Spring: 10–11.

Leal Filho, W. D. S. (1992) *A Floresta Amazonica*, Paris: UNESCO.

Leal Filho, W. (1994a) 'Environmental awareness and tourism in Portugal', *Sustainable Tourism* XII: 1–14.

Leal Filho, W. (ed.) (1994b) *Environmental Education in Small Island Developing States*, Vancouver: The Commonwealth of Learning.

Leal Filho, W. D. S. and Hale, M. (eds) (1994) *Trends in Environmental Education Worldwide*, London: London Guildhall University.

Leal Filho, W. (1996a) 'Trends in environmental education in Europe', *Journal of Environmental Education* 27 (3): 1–8.

Leal Filho, W. (1996b) 'Eurosurvey', in Harrisson, G. and Blackwell, C. (eds) *Environmental Issues in Education*, Aldegate: Arena.

Leal Filho, W. (1997) 'Implementing Agenda 21: shared problems and perspectives of co-operation', *Proceedings of the European Workshop on Agenda 21*, Frankfurt: Deutsches Institut für Erwachsenenbildung.

Local Government Management Board (1995) *Local Agenda 21 Survey 1994/1995*, London: LGMB.

Organisation for Economic Co-operation and Development (1993) Environmental Policies and Industrial Competitiveness, Paris: OECD.

Pirages, D. C. and Elrich, P. R. (1974) *Ark II: Social Response to Environmental Imperatives*, San Francisco: Freeman.

Schleicher, K. (1996) *Umweltbewusstsein- und Umweltbilding in der Europäischer Union*, Hamburg: Krämer Verlag.

Schmidheiny, S. (1992) *Changing Course: a Global Perspective on Development and the*

Environment, Cambridge, MA.: MIT Press.

Templet, P. H. (1995) 'Equity and sustainability, an empirical analysis', *Society and Natural Resources*, 8: 509–23.

Thioune, O. (1993) 'An Analysis of Environmental Education Techniques Suitable for Rural Areas in Senegal', unpublished Postdoctoral thesis, University of Bradford.

United Nations (1992) *The UN Conference on Environment and Development: A Guide to Agenda 21*, Geneva: UN Publications Office.

UN Environment Programme (1995) *The Role of Indicators in Decision Making*, Nairobi: UNEP.

World Bank (1994) *World Development Report*, New York: Oxford University Press.

World Commission on Environment and Development (1987) *Our Common Future*, New York: Oxford University Press.

World Health Organisation (1995) *Concern for Europe's Tomorrow*, Stuttgart: Wissenschaftliche Verlagsgesellschaft.

4 Beyond volunteering and rhetoric

Implications of Local Agenda 21 initiatives in Wales

Michael Clark and Alan Netherwood

Introduction

Local Agenda 21 invites contradictory interpretations. It can be argued that it has made a real contribution to the objectives of sustainable development by encouraging numerous worthwhile initiatives and changes. Alternatively, it may be seen as a weak apology for lack of official commitment, a misleading distraction and a substitute for necessary action. We may value LA21 as a sustainability flagship that has helped the process of environmental education and achieved much through mass participation, voluntary effort and the leading work of charities and local authorities, or dismiss it as a haphazard jumble of under-funded, often ineffective, substitutes for real environmental policy, regulation where it matters and genuine local ownership of community initiatives. Our discussion in this chapter gives particular attention to recent events in Wales, and to the criticism that LA21 is little more than politically expedient rhetoric. We question the leading role of local authorities, and explore some of the challenges, difficulties and achievements of growing official reliance on work contracted out to NGOs, and often carried out by unpaid or trainee volunteers.

Volunteering has played an important part in urban nature conservation, and its extension to encompass wider environmental and sustainability objectives includes significant inputs to LA21 (Marshall and Patterson, 1996: 323, 327). Dependence on volunteer effort, labour or management raises a number of issues which are usually addressed from the perspective of the (UK) National Council for Voluntary Organisations (NCVO). Here the most significant voluntary sector activities are social care (37 per cent), accommodation and housing (19 per cent) and culture and recreation (14 per cent) (Joseph Rowntree Trust, 1996: 2). The (independent) Commission on the Future of the Voluntary Sector in England, set up by NCVO, includes the following in its agenda for future action:

- safeguard the independence and diversity of voluntary and community organisations;
- harness the enthusiasm and commitment of young people;
- formalise relationship with government: agree explicit requirements and priorities;

- 'voluntary work must not substitute for activity that is properly the responsibility of the state or the market';

and has six basic principles:

- unique qualities of voluntary action to be recognised by public policy;
- partnership on an equal basis;
- the role of users (as stakeholders) is crucial;
- freedom to act as advocates;
- professional management not to deflect from the sector's purpose and aims;
- diverse funding sources to guarantee independence.

(adapted from Joseph Rowntree Trust, 1996: 1–2)

The importance of voluntary effort in recent environmental and sustainability initiatives reinforces the NCVO case for recognition, but it is far from clear if the agenda or principles outlined above fit recent practice, or if it is realistic to expect their adoption by the cash-strapped and opportunistic authorities and other bodies that have come to rely on the voluntary sector. This reliance can include the initiatives necessary to implement, or at least to be seen to be working towards, LA21.

A leading role for local government?

The deadline, agreed by governments at the United Nations Conference on Environment and Development in Rio in 1992, was something of a shock to many of those it committed to meeting LA21 commitments by the end of December 1996.

> . . . an alliance of international municipal bodies determined to get local government written into the Rio outcome and lobbied accordingly . . . Disappointingly, the only mention of local government and local communities in the Rio Declaration on Environment and Development, in Principle 22, is unsatisfactory. But the far greater prize of a separate chapter in Agenda 21 was achieved. Chapter 28 (Agenda 21's shortest) recognises that because so many of the problems and solutions addressed by Agenda 21 have their roots in local activities, the participation and co-operation of local communities will be a determining factor in fulfilling its objectives. Local government's role is recognised both as planner and implementer of relevant policies and as educator and mobiliser of local public opinion . . . there will be a dialogue between each local authority and its citizens, local organisations and local business which would lead to the adoption of a 'LA21'. The importance of consultation, consensus-building and awareness-creation are emphasised. A deadline of 1996 is given for the completion of this process, although with the stipulation that this covers 'most local authorities' rather than all.

> Current estimates suggest that local authorities will be responsible for implementing about 40 per cent of the EC's Fifth Action Programme and that over two-thirds of Agenda 21 commitments cannot be delivered without the commitment and co-operation of local government . . . The emphasis is more on local and community-based action taking place within an enabling national framework than on traditional 'top-down' approaches which put the national state centre stage and ignore or downgrade the role of other institutions and groups.
>
> (Gordon, 1994: 138, 148)

Local authorities had to find ways of interpreting the Agenda's worthy and very general objectives, and of turning these into practical action. This was not helped by the contradiction at the very core of sustainable development. If development is (wrongly) interpreted to mean growth, how can economic expansion be sustained against a fixed, and degrading, natural resource base? Also there is confusion between environmental objectives and some of the more contentious, politically difficult, matters raised by the drive for sustainability. Should we dismiss calls for the routine incorporation of sustainability objectives as a weak mix of reformist zeal and emotive popularism? Or should we look to initiatives such as LA21 as a means, however flawed, of achieving more responsible human behaviour?

Development requires social justice as well as environmental sensitivity. How might LA21 involve the victims of poverty and environmental degradation? It is intended to ensure full participation of groups which are under-represented in most official decision making: women, young people and members of disadvantaged communities. Yet responsibility for implementing LA21 has fallen largely on local government. Their officials and elected representatives are well placed to give the strategic overview and general policy context for local sustainability initiatives. But, they also tend to represent those interests and professions that have brought us to the present state of affairs. They may also have been superseded by new, less consistent (and much more widely cast) forms of local governance, matching (and perhaps explaining) less effective post-Fordist modes of regulation (Goodwin and Painter, 1996: 635, 636–7; Patterson and Theobald, this volume, Chapter 12).

Whether or not local government accepts the status quo, or is actively involved in trying to change it – for instance by acting to eradicate homelessness, alleviate poverty, or achieve environmental improvement – its capabilities are severely constrained. This may be due to it representing interests (landowners, capitalists, managers, professionals in comfortable jobs . . .) that fear anything that smacks of Socialism. Or, it may be that local government simply cannot deliver. Either it isn't competent, or it isn't permitted to be because its powers and its funds have been cut.

> Whitehall policies over the past decade have forced local government into a very short time-frame governed by strict financial constraints. Key concerns all too often centre on which local services are to be 'saved' and which cut,

> and by how much to meet externally imposed financial ceilings. Long-term overall perspectives have largely gone by the board.
>
> (Gordon, 1994: 149)

Voluntary action: complement or substitute?

In the UK this hollowing out of the local (democratic) state (Patterson and Theobald, 1995: 5) can be traced back to the increasing role central government has given quangos and other appointed bodies and commissions over the last 75 years, and recently seen in a raft of hostile initiatives: compulsory competitive tendering, capping, tight spending limits and difficult performance targets. These have favoured voluntarism – relying on people's good nature and consciences rather than regulation or economic instruments, and expecting that necessary, virtuous work will be done for free.

There is tension between the advantages that volunteers can bring, and the risks of depending on cheap labour 'to fill gaps in the labour force' (Department of the Environment, 1972: 24). 'The success of self-help activities often depends on people working without pay. Beyond a certain point, however, reliance on donated labour can be just another means of exploitation, and is bad economics as well' (Stokes, 1981: 134).

Advantages listed by the Department of the Environment Working Party report on the Role of Voluntary Organisations and Youth in the Environment, prepared for the June 1972 UN Conference on the Human Environment in Stockholm, include:

> 1 Where volunteers supply a service for a need which government has so far not recognised.
> 2 Where volunteers put pressure on government policy.
> 3 Where volunteers carry out a particular function because they do it as well or better than government.
> 4 Where volunteers enjoy some recreational or educational experience because of the work.
>
> (Department of the Environment, 1972: 25)

A quarter of a century later the educational and community benefits of volunteering may be seen to go way beyond these final two categories. This is partly because the voluntary sector's contribution has tended to be overlooked. Beckford reports that the Wolfenden Report of 1978:

> concluded that 'the Voluntary system . . . can best be seen in terms of the ways in which it complements, supplements, extends and influences the informal and statutory systems' (Wolfenden Committee 1978: 26). The expectation was that the voluntary system would *extend* the scope of statutory services by identifying new needs and pioneering new methods; *supplement*

> the state's services by providing alternatives and by attracting the help of volunteers who would not be attracted to work in the statutory sector; and *influence* the public sector by showing how improvements would be made and by criticising statutory arrangements. But . . .[this] seriously underestimated [the] importance of voluntary initiatives . . . [the] sole providers of certain services (for example, marriage guidance, citizens' advice bureaus, emergency help telephone lines, life boats, animal welfare, first aid in public places).
>
> (Beckford, 1991: 38)

Subsequently, relationships between voluntary agencies and the state have become more contractual, and less based on trust. 'Many voluntary organisations increasingly operated as clients of the state' (Beckford, 1991: 39–40), under conditions where funding was uncertain, and there was a certain loss of independence.

Paralleling, and to some extent contrasting with the voluntary sector's incorporation within state welfare provision, a form of participatory democracy has emerged which stresses self-help and ordinary peoples' ability to overcome adversity, if they are allowed the means (and if government, landowners and other competing interests get off their back). Anarchist traditions of mutual aid (Kropotkin, 1913), self-help through guilds, friendly societies and building societies (Davis Smith, 1995: 28–35) and the challenge of 'new economics' re-emerge in calls to awaken community resourcefulness (Dauncey, 1988: 91), acknowledge the full extent and importance of community and domestic work and help people help themselves (Stokes, 1981).

Paradoxically, many of these sentiments are shared by those who look to charities and voluntary work as ideologically preferable to state provision. The practicalities of mobilising community resources to meet community needs (Chalker *et al.*, 1979) may draw on traditions hostile to the Welfare State, and opposed to public-sector provision of responsibilities which are believed to lie with, and be best served by, charities or the family. Here much may be made of the religious basis of philanthropy, and of the way that 'managerialism' changes charities, arguably for the worse, as they are taken over by 'contract culture', and by full-time professionals (Whelan, 1996). Others argue that voluntary organisations are not, and should not be treated as, different from other types of business, and that they require modern management techniques as the state is no longer willing, or (at a local scale) permitted to take responsibility for a variety of worsening problems. From this perspective, it is anachronistic that in an increasingly secular society, so much voluntary work is provided by religious groups. The challenge is to ensure the provision of quality service by following good working practice (Gann, 1996: 2–3, 73). Managerialism and the contract culture appear to have gone furthest in voluntary bodies dealing with social welfare, but the sector's complexity and scale makes it unsafe to generalise. Some environmental organisations correspond with the management situations which Gann describes and prescribes, while others fit the notion of self-help.

These circumstances have led to the development of academic disciplines:

'non-profit studies' and 'voluntary sector studies' (Davis Smith *et al.*, 1995: 3), and to a great variety of responses within the voluntary sector:

> Voluntary bodies have received a substantial increase in direct funding from government and the opportunity to compete for new service delivery tasks. But in doing so, they are entering a new universe with very different rules of engagement: in shorthand, the 'contract culture'. Some are well equipped to cope with it. Most large voluntary agencies have taken on board the lessons of the management revolution of the 1980s and kitted themselves out with all the paraphernalia of the enterprise culture: mission statements, logos, personal identification with tasks, 'passion' (even obsession) for excellence. Others resisted but have recognised that survival has meant being able to play the game according to new rules. Others are still simply bewildered or hardly aware of what the rules are – 'generic' organisations operating at community level and many ethnic and women's groups. Others still have been deliberately excluded because their objectives do not mesh with the project: these are the deplorable 'pressure groups' – mainly advocacy and campaigning bodies.
> (Deakin, 1995: 62)

In addition to the retreat of the state (Deakin, 1995: 63), and various ideological positions which favoured self-reliance, we should acknowledge the institutional adoption of a period of 'volunteer' activity as part of education, training or professional socialisation. Practice varies, but it has become usual for vocational or applied courses to include a work placement, and for a career in conservation, environmental management or development to be preceded by a period of volunteer activity. As with a gap year, usually taken before entering higher education, and involving a mixture of travel and worthwhile activity, it is difficult to criticise this level of commitment, and the personal development which it achieves. But there are risks and costs involved which mean that such opportunities are not universally available. It may be useful to curtail the career opportunities of those who lack the motivation, coping strategies and interpersonal skills necessary to make the most of an extended and demanding period of volunteer work or placement. But it is less safe to encourage a macho culture, or for access to private means or favourable family circumstances to distinguish volunteers from their contemporaries. Perhaps the biggest risk is of exploiting donated labour, and of this work not being up to standard.

Returnees to employment may find that a period of unpaid work is necessary to re-establish their professional credentials, and many retired people find it rewarding to be involved in charitable or community-based activity. The unemployed receive contradictory messages: benefits such as the Jobseeker's Allowance require them to be available for work at any time, and at short notice. Payment for voluntary work can lead to punitive loss of unemployment or invalidity benefit (Blacksell and Phillips, 1994: 49). Yet one way that the unemployment figures have been reduced has been through training schemes which involve working alongside volunteers, or on projects organised by the voluntary sector. With such a variety of motivations, it is dangerous to make generalisations, or to assume that donated labour is either problematic or advantageous.

At its most extreme, a culture of volunteering may rely on a few (paid) managers overseeing large numbers of volunteers, often short term or part time. This can deny conventional jobs, and be less efficient. It may also undermine professional or specialist careers and salaries, and so affect performance, recruitment and retention in the long term. Litter clearing, school library provision and classroom assistance, care services, conservation and environmental management illustrate the risk of volunteers *substituting* for established posts, though, paradoxically, these may offer some of the best opportunities for worthwhile community involvement and service. The trick seems to be to encourage extra work outside the money economy, without allowing the new initiatives to become a way of saving money. In other words, adopt the European Community's *additionality principle*, which was intended to ensure that Member States did not use Community funds to reduce spending from the national exchequer.

The voluntary sector: real work?

Not-for-profit activity in the UK involves as many people as the paid economy (23 million). On the basis of time contributed, notionally paid at the average 1993 hourly rate of £7.83, volunteers' contribution to the national economy has been estimated at £41 billion, the third largest component of Gross Domestic Product (Joseph Rowntree Trust, 1997: 2). Three million voluntary committee members run an estimated 230,000 to 300,000 voluntary organisations in England and Wales, and employ a quarter of a million people (Davis Smith *et al.*, 1995: 2). These statistics may be seen as misleading because the voluntary sector encompasses everything from unpaid or paying volunteers, through underpaid workers for charities and on 'training schemes' and community projects under the professional, paid, management of NGOs and charities (Blacksell and Phillips, 1994; Leavesley, 1995: 310–11) through to large housing associations, the National Trust and other big, non-profit-making bodies. It includes sporting activities, youth groups and organisations, and some campaigning and interest groups: though Charity Commission rules prohibit direct political involvement. Many people will be members of a number of different organisations, and will join for a wide variety of reasons. Although official statistics suggest less participation than some global estimates, the rate of growth has been dramatic.

Funding and other difficulties have prompted UK local authorities to rely on this 'third force' of charities, volunteers and various types of NGOs. Increasingly, these have been encouraged to bid for contracts, and commit themselves to service agreements and specifications, rather than apply for grants (Lewis, 1996: 99). This process has been favoured by the ideology of compulsory competitive tendering: the 'contracting out' of local government services; and a 'contract culture' which favours the greater discipline of legally binding service agreements and contracts in place of grants (Adirondack and Macfarlane, 1993: 5).

Table 4.1 Membership of selected voluntary organisations, UK

	Thousands	
	1971	*1994*
National Trust and National Trust for Scotland	315	2,454
Royal Society for the Protection of Birds	98	870
Greenpeace		300
Wildlife trusts	64	263
Civic Trust	214	222
World Wide Fund for Nature	12	187
Woodland Trust		170
Friends of the Earth	1	112
Ramblers Association	22	100

Source: Data selected from Central Statistical Office, *Social Trends* (1995), table 11.4.

> Central government support for the voluntary sector totalled £3.6 billion in 1993–94. And voluntary sector income from UK local authorities was nearly £1.3bn in 1994–95, a 12 per cent increase on the previous year . . . many charities complain that they are increasingly substituting for statutory provision, rather than complementing it.
>
> (Bibby, 1996)

While the voluntary or non-profit sectors' members, employees and volunteers are generally enthusiastic, highly committed and often have considerable professional and practical skill in their particular field, resources are usually limited. Support for initiatives such as those emerging from LA21 may only be possible if volunteers' costs can be met. This has created the interesting situation where charitable bodies compete for contracts to supply local authorities and other official agencies with expertise, including the delivery and management of environmental and community-based schemes. There are obvious risks here. Concern to win subsequent contracts may push voluntary bodies into consultancy type roles, and into a semi-commercial status in which delivery of a product acceptable to the budget holder may become more important than something which reflects the priorities of the community which is the nominal focus for intervention. There is a risk that council bureaucrats will merely be replaced by similar people working for an NGO, with no increase in community representation, participation or involvement, and little of the 'ownership' implied by words such as enabling and facilitating. Finally, councils may be reluctant to relinquish control of initiatives intended for 'community' ownership. The current development of LA21 initiatives in Wales illustrates and challenges a number of these points.

LA21 and the voluntary sector in Wales

Many of the Welsh local authorities subject to reorganisation in April 1996 have had to begin the whole process of Agenda 21 from scratch. While in some areas this has meant a reinvention of the wheel and much covering of old ground, in others the reorganisation process has amalgamated expertise, and has enabled officers and councillors to learn from the mistakes of previous attempts at involving communities in sustainable development. A number of Welsh authorities are promoting the fact that they are addressing internal issues first, by undertaking the Environmental Management and Audit Scheme for Local Government (EMAS):[1] perhaps best described as a variant of the, more familiar, business-orientated BS 7750, EU Eco Management Initiative (also called EMAS) and the relatively new ISO 14001. Levett states that the Eco-Management and Audit Scheme for UK Local Government, used in some form by rather less than half of UK local authorities by June 1996 'offers a systematic framework for a local authority to state its environmental aims, identify how it affects the environment, decide what actions to take and who will take them, and then monitor and report on progress' (Levett, 1996: 329).

While this in itself is laudable, as corporate initiatives being undertaken by local government are necessary to provide the basis for future sustainable policy development, it could be argued that authorities are skirting the thorny issue of community consultation: watching while other more proactive authorities make mistakes, or gain experience in the community. Communities are left disempowered from taking action five years on from Rio, due to reorganisation and the local authorities fear or mistrust of participative processes.

However, in reality communities are just getting on with it, with or without local government help. This mixture of locally based pragmatism and expedient use of the voluntary sector contradicts the assumption that LA21 depends on established local government institutions and procedures. Perhaps the real rhetoric of LA21, in the sense of implied insincerity, or exaggeration (Fowler and Fowler, 1964: 1070), is that local authorities have such an indispensable role in the process. Indeed, in how many of the plethora of case studies cited in the literature do LA21 consultations involve people from actual communities (residents' associations, community councils and individuals) rather than interested agencies and traditional environmental activists? This begs further research, and we suggest that a good starting point is local authorities' apparent reluctance to permit, let alone support, rival forms of representation and different priorities. These alone may account for much of the difficulty in implementing LA21, especially the tendency to interpret it in narrow ways which correspond to established practice, and which adopt existing schemes (usually from the voluntary sector) rather than risk major commitments of their own resources or the wider implications of a thorough review of existing policies and priorities.

Many of the LA21 consultations undertaken by local authorities just add to the corporate rhetoric of local government approaches to sustainable development, and do little to facilitate practical action which helps communities address their environmental, social and health issues. However, in Wales, community agencies and NGOs are taking part in more practical ways of helping active communities to implement LA21 with relatively little help from local authorities. This adds an extra dimension to discussion about the value and limitations of voluntary sector input. NGOs and their volunteers provide additional community-based initiatives, broadly linked to local interpretations of LA21. Paradoxically, given the damaging effects of local government's reduced funding and curtailed powers, much of the support for these projects has come from central government. Environment Wales, a Welsh Office initiative, funds Development Officer posts in the key environmental volunteering agencies – BTCV, Community Service Volunteers, Keep Wales Tidy, Groundwork Cymru, the Welsh Wildlife Trusts, The National Trust and RSPB. As well as promoting environmental volunteering, these organisations have traditionally provided advice and support to communities on funding and practical techniques for environmental projects, and provided them with access to both local and national networks. The Development Officer posts support these functions and also give communities access to Welsh Office funding for environmental and sustainable development projects through Environment Wales.

The range of activities is huge, enabling communities and groups to get involved in biodiversity projects, countryside and habitat management, community recycling, urban wildlife, sustainable transport, landscape improvements and environmental awareness-raising activities. Importantly, the initiative and its partners aim to involve under-represented groups and individuals in sustainable development – women, young people, those with special needs and people from disadvantaged communities. The activity of these NGOs not only provides an 'apolitical' input into local sustainable development, but also gives valuable information and feedback for local authorities carrying out local development planning in Wales.

Some Welsh local authorities are realising the potential for working in partnership on LA21 – with large-scale collaboration with Groundwork Trusts, partnerships in habitat and woodland management with BTCV and the Welsh Wildlife Trusts, and neighbourhood-level appraisal initiatives through Community Service Volunteers and Keep Wales Tidy. However, the majority of Environment Wales partners' work involves individual community projects that local authorities would find it very difficult to access and support. This is a tier of LA21 activity that is seldom acknowledged in the literature, and which enables citizens to have a role in the decision-making processes of local sustainable development. Perhaps this explains the reluctance of some local authorities in Wales to give up control of LA21 to the communities they purport to serve. What is needed in Wales, and elsewhere, is for local authorities to create a balance between the corporate rhetoric of LA21 which means so little to the average person in the street, and practical projects which would positively affect the quality of life of that person. They can do this through partnership in a relatively informal way (and in some cases are doing so) by providing matching funding for community groups to apply for lottery or Environment Wales grants, and to support the Environment Wales partners in their essential role of facilitating local sustainable development. However, this may prove to be difficult as resource constraints in Welsh local government have been particularly acute since the latest round of local government reorganisation, and political interests may counter any moves to democratise local sustainable development.

Less rhetoric: more practice

All these doubts suggest that a less rhetorical interpretation of sustainability is required for it to realise its potential. It must be supported by adequate resources and powers if it is to be anything other than a cynical exercise in media manipulation and feel-good propaganda. But LA21, like every other top-down initiative intended to energise a troubled, but passive, local population is prone to political manipulation of all forms. In some ways the target populations remain part of the problem. So complete empowerment may not just be unpredictable, and politically unacceptable because it is out of the control of local bureaucrats and politicians, it may even be counter-productive. However, the local people are also, and must also be seen as, a necessary part of any long-term solution. Without their willing and active participation, few initiatives survive for long. This brings LA21 down to earth, along with every initiative to sort out localised problems of crime, dereliction, under-attainment and ill health. All rely on education, and on a form of political mobilisation. These are necessarily diverse, even chaotic. So we should not be ashamed by the diversity of responses to LA21, or too anxious that it has not worked, or had the same effect, everywhere. It is about Development in the sense of Education, and that is a process which defies tidy answers.

Conclusions

Local Agenda 21 is important because it exposes our institutions and assumptions to the possibility that:

- The environmentally disasterous and regressive effects of more than 500 years of expropriation and degradation (the long process of enclosure, modernisation or 'development') may be overturned, without revolutionary change or catastrophic 'adjustment', through such simple expedients as ensuring that externalities are internalised, regulation is effective and consequences are acknowledged, assessed and acted upon.
- Government, at all levels, should do more to acknowledge the role and potential of non-governmental and voluntary-sector organisations.
- Public agencies and government bodies should provide community groups with matching funds to maximise the potential of competitive funding from such sources as Lottery or Environment Wales grants, rather than seek to 'lever in' additional private sector resources (Evans, 1996, i).
- People, their communities and the environments on which they depend, and which they may threaten, should be recognised as the measure by which government should be judged, and as an important component in its decisions and its actions: the means and the ends.
- Environmental initiatives can help mobilise people and their communities, and inform institutional relationships and professional attitudes, in ways which amount to a far more purposeful and deep-seated education than their scale, cost or immediate outcomes suggest.

- The environmental/'green' focus of most LA21 initiatives to date misses the more fundamental questions raised by Rio. Any interpretation of sustainable development which fails to question, and act on, injustice and poverty, and their human and environmental consequences, misses its most important implications, and likely benefits. Not because these are explicit in its resolutions and treaty (by and large these reflect the status quo), but because the resulting actions will help mobilise, educate and sensitise enormous numbers of individuals and groups. Their dissatisfaction, aspirations and abilities could overcome much that is wrong.

In short, LA21 should be approached from the broadest possible perspective of environmental education, and further work is necessary to establish the contribution which this makes to development in a qualitative and absolute sense. There is considerable scope for environmental management specialists to contribute to, and learn from, research into the role, potential and problems of the voluntary sector, and there is a need to monitor how volunteers are used in environmental and sustainability initiatives. Institutional and behavioural blocks to local initiative should be identified, and the means found to overcome them. We need to question the leading role which local authorities have assumed in the implementation of LA21. And we should also question the financial and other limits which central government has put on local democracy. Some established practice contradicts principles of additionality and may be counter-productive, for example, by substituting token or badly funded activities for real projects, by failing to reward certain trades and specialisms adequately or provide realistic career paths other than in 'management', or by so compartmentalising LA21 that its application is restricted to a few, marginal, initiatives. LA21 is a sham and a smokescreen if it does not facilitate the more effective questioning of plans and policies on sustainability grounds, but the process of which it is a part necessarily goes beyond institutions and procedures. We should look for its effects at the local level, as people find better ways of expressing their concerns and establishing priorities, overcoming immediate difficulties and safeguarding those aspects of their environment or way of life which they value. Participation, self-help, NGOs and volunteers may prove to be more sympathetic and effective than 'cadres' and 'experts', but there is also likely to be a role for the 'facilitator', specialist community development officer (who may also carry a health, environmental or education remit) and local government employee. From a negative perspective, the scope for initiatives led by 'new' social actors implies some failure within established political parties, local bureaucracies and other organisations, so an alternative may be more effective communications within, and mobilisation by, the existing order. At the larger scale, such internalisation and routinisation are necessary for progress towards sustainability goals. Whether this is satisfactory depends on the quality of environmental education and scientific understanding, and the extent to which these can escape the restrictions imposed by vested interests and the status quo.

Note

1 This is an evaluative framework developed by LGMB and others, on the basis of the European Union's EMAS scheme for industry. It is a voluntary, rather than a compulsory, scheme which, after being piloted in seven UK local authorities, many others have chosen to enter. Its aim is to describe and quantify an authority's direct and service effects on the environment, on a corporate basis, and to help it continuously to improve on its current performance . . . the scheme does not prescribe particular environmental policies or performance standards, but establishes a management framework to instil good and continuously improving practice (i.e. it aims to provide quality assurance rather than quality control).

(Selman, 1996, 93)

References

Adirondack, S. and Macfarlane, R. (1993), *Getting Ready for Contracts: a Guide for Voluntary Organisations* (2nd edn), London: Directory of Social Change.

Bibby, A. (1996), 'Farewell coffee morning, hello authority cash', *Observer*, 27th October.

Blacksell, S. and Phillips, D. (1994), *Paid to Volunteer*, London: Volunteer Centre UK.

Chalker, L., Morris, S. and Rowe, A. (1979), *We Are Richer than We Think: How to Mobilise Community Resources to Meet Community Needs*, London: Community Affairs Department, Conservative Central Office.

Dauncey, G. (1988), *After the Crash: the Emergence of the Rainbow Economy*, Basingstoke: Greenprint.

Davis Smith, J. (1995), 'The voluntary tradition: philanthropy and self-help in Britain 1500–1945', in Davis Smith, J., Rochester, C. and Hedley, R. (eds), *An Introduction to the Voluntary Sector*, London: Routledge.

Davis Smith, J., Rochester, C. and Hedley, R. (1995), An Introduction to the Voluntary Sector, London: Routledge.

Deakin, N. (1995), 'The perils of partnership. The voluntary sector and the state, 1945–1992', in Davis Smith, J., Rochester, C. and Hedley, R. (eds), *An Introduction to the Voluntary Sector*, London: Routledge.

Department of the Environment (1972), *50 Million Volunteers: a Report on the Role of Voluntary Organisations and Youth in the Environment*, London: HMSO.

Evans, J. (1996), 'Forward by the Welsh Office', *Environment Wales Annual Report 1995–1996*, Cardiff: Welsh Office.

Fowler, H.W. and Fowler, F.G. (eds) (1964), *The Concise Oxford Dictionary of Current English*, Oxford: Clarendon Press.

Gann, N. (1996), *Managing Change in Voluntary Organisations: a Guide to Practice*, Buckingham: Open University Press.

Goodwin, M. and Painter, J. (1996), 'Local governance, the crisis of fordism and the changing geographies of regulation', *Transactions of the Institute of British Geographers*, 21: 4, 635–665.

Gordon, J. (1994), 'Letting the genie out: local government and UNCED', in Thomas, C. (ed.), *Rio: Unravelling the Consequences*, Ilford: Frank Cass.

Joseph Rowntree Trust (1996), 'The future of the voluntary sector', *Findings; Social Policy Summary 9*, July, http: //www.jrf.org.uk/social_policy/SP9.html.

Joseph Rowntree Trust (1997), 'The economic equation of volunteering', *Findings; Social Policy Research 110*, February, http: //www.jrf.org.uk/social_policy/SP110.html.

Kropotkin, P. (1913), *The Conquest of Bread*, Anarchist Pocketbooks 4.
Leavesley, J. (ed.) (1995), *Occupations 1995*, Bristol: Careers and Occupation Information Centre.
Levett, R. (1996), 'From eco-management and audit (EMAS) to sustainability management and audit (SMAS)', *Local Environment*, 1: 3, 329–334.
Lewis, J. (1996), 'What does contracting do to voluntary agencies', in Billis, D. and Harris, M. (eds), *Voluntary Agencies: Challenges of Organisation and Management*, Basingstoke: Macmillan.
Marshall, R. and Patterson, J. (1996), 'The role of the voluntary sector in urban nature conservation in Britain, *Local Environment*, 1: 3, 323–328.
Patterson, A. and Theobald, K.S. (1995), 'Public sector restructuring and local environmental practices', paper presented at the Annual Conference of the Institute of British Geographers, University of Northumbria, Newcastle, 3rd January, mimeo.
Selman, P. (1996), *Sustainable Development: Managing and Planning Ecologically Sound Places*, London: Paul Chapman.
Stokes, B. (1981), *Helping Ourselves: Local Solutions to Global Problems*, London: Worldwatch Institute / Norton.
Whelan, R. (1996), *The Corrosion of Charity: From Moral Renewal to Contract Culture*, London: Institute of Economic Affairs, Choice in Welfare Series No. 29.

5 Hopes and fears for Local Agenda 21

Ken Webster

Introduction

It's the rich what gets the pleasure.
It's the poor that has the pain.
It's the same the whole world over.
Ain't it a bloody shame.

(Music Hall song written by Bob Weston and Bert Lee 1930)

The World Wide Fund for Nature UK's education department comprises seven specialist units covering formal and non-formal education. This chapter is written from the perspective of the community and governance unit, that since the Rio Earth Summit has concentrated much of its energy on supporting the key audiences within Local Agenda 21. At the time of writing this chapter WWF UK has produced more than 20 LA21 resources and training programmes.

This chapter is a commentary on, and an overview of some of the issues regarding the processes within LA21 and the people who have driven it during its evolution since the Rio Earth Summit in 1992, particularly in relation to WWF UK's own experience. It attempts to describe how the Agenda is moving from purely green to one that is increasingly demonstrating the interrelationship between the environmental, social and economic elements of sustainable development. It provides a broad-stroke approach to the political dynamics and some of the difficulties of overcoming the inertia of the status quo. There is no pretence that the story of LA21 so far, is a smooth and logical step-by-step progression – it is throwing up too many challenges for that. In the examples of work used as illustrations there is real optimism about our ability to do things better, particularly in the way communities are addressing some new and some familiar concerns about their local quality of life issues. But also, there is real anxiety in our awareness that LA21/Agenda 21 is not the only world wide agenda. The Multilateral Agreement on Investment (MAI) for example, if finally implemented in anything like its early drafts, poses a major potential threat to all that LA21 will have achieved by local authorities and communities. The *between-the-lines* question in this chapter asks if we have yet been made strong enough by the LA21 process to stand up to the negative and retrogressive forces that continue to perpetuate our present unsustainable state. Just as different levels of governance, community and business are beginning a new dialogue for sustainability, those who represent the advantaged and the status quo are wielding massive economic power, through MAI, to protect their position. The question has to remain only partly answered in this chapter because we are dealing with complex, long-term and real-life issues that to a great extent locate us at the crossroads in a journey which will determine whether the world turns towards or away from sustainability.

Setting the scene

Since 1992 a growing number of organisations have taken up the implicit challenge in Nitin Desai's (Under Secretary-General for Policy Co-ordination and sustainable development) description of Agenda 21 as 'a set of integrated strategies and detailed programmes to halt and reverse the effects of the environmental degradation and to promote environmentally sound and sustainable development in all countries' (Desai, 1993: i). Different people and organisations have focused on the chapters of Agenda 21 that have greatest bearing on their area of work. Chapter 28 has a particular resonance for local authorities and communities, and Chapter 30 for business and industry. These are the areas on which WWF UK's LA21 strategy has focused. One key sentence in Chapter 28.2 (a) says that, 'By 1996, most local authorities in each country should have undertaken a consultative process with their population and achieved a consensus on a Local Agenda 21 for the community'. Those three Cs: consultative, consensus and community, are a Trojan Horse in which are hidden many unknowns. These are real challenges to the way that local and national governments relate to their communities, and vice versa. There are real issues of empowerment, ownership and participation, and at last (at least in relation to the relationship between government and community) it seems it is being recognised that top-down and set-in-concrete solutions, that are presented to communities as a *fait-accompli*, do not equate to effective consultation. It is also being learned that methods of consultation which present and defend are incapable of achieving consensus or ownership about anything. During the first two months of 1998 the UK government invited national consultation on sustainable development, life-long learning and modernising local government. Though it is easy to pick fault with it as a flawed process, at least the starting point of the consultation was a series of questions rather than answers that had already been agreed behind closed doors.

LA21 is essentially an exhortation for action to achieve sustainable development at all levels and all areas of society. However, responding to the exhortation is proving quite demanding because the successful and full implementation of LA21 is making demands for a new understanding of the dynamics of decision making. Politicians at all levels who have been brought up in the adversarial model of parliamentary debate are beginning to realise its limitations in achieving consensus and lasting change. In WWF UK's first four Interactive Seminars, politicians, officers and community representatives from 35 local authorities have begun the process of exploring the value and benefits of community participation in sustainability policy development. Implicit within this work is the shift from purely representative democracy to a balance between representative and participative democracy. A sustainable socio-economic system will require increased political accountability, improved decision making and increased community ownership of local policies that affect social integrity, environmental protection and economic development. Because sustainable development is affected by the daily decisions people take regarding their life style expectations it is essential that there is a supporting process of life-long learning through which the implications of those decisions can be explored and understood.

The learning process in the community is about creating the motivation in consumers to distinguish between wanting and needing, and about knowing the boundaries which mark the cross-over point from essential to non-essential consumption. The learning consumer is provided, through programmes such as the joint Global Action Plan (GAP) and WWF UK Action at Home programme, with the information and methodology to examine their domestic impact on the environment, and to adopt new approaches that minimise that impact. But it is not just about the individual. At a community or collective level it is essential that local people are able to become fully involved in the process of identifying and shaping responses to their concerns and issues, visions and plans. LA21 is a mechanism by which organisations, individuals and communities can together learn, plan and implement a greater degree of collective responsibility for the environment. Community groups in the WWF UK and Reading Borough Council's GLOBE (Go Local On a Better Environment) programme have spent four years so far developing neighbourhood action plans that show both time scale, and what the community and the local authority are individually best placed to deliver. This mature sharing of responsibilities demonstrates a major step away from the traditional confrontational relationship that characterises so many local situations. The concept of the learning community is important for individuals, communities and organisations. It is as important for those people running business and political/bureaucratic organisations to also engage in a learning experience where the curriculum is based on a continuum that starts with local and passes through national to international perspectives. The context and clear purpose of the learning is to enable people to make informed and equitable life style decisions regarding their impact on their own daily lives and that of others.

WWF UK and Local Agenda 21

WWF UK's vision recognises that the greatest threat to species and habitats comes directly or indirectly from people, making it necessary to reconcile the needs and aspirations of people with the conservation of nature. In pursuing this vision, WWF UK has for many years made a considerable investment in Education for Sustainability. The rationale for this is found in WWF UK's education policy, which states:

> If the WWF mission is to be achieved then the majority of people within

> society will need to contribute to the development of a social context where sustainable and careful resource use are given in terms of the choices they make at home, in the community, in the world of work and in the high street.
>
> (Peter Martin, Head of Education WWF UK, 1996/7: unpublished internal document)

Within this context the WWF UK education goal is:

> to ensure that key areas of society that have the potential and legitimacy to influence and implement the lifelong learning process, are supported by: central governmental policy; adequate and appropriate training with resources/materials; informed capacity within the elected membership and executive; public opinion and professional acceptance.
>
> (ibid.)

The guiding principle underpinning WWF's LA21 strategy is a commitment to the notion that sustainable development can only be achieved when all stakeholders are able to fully participate in the processes of informed decision making. What this means in practice is that people, having articulated their vision for life in their area over the next five, ten, or twenty years, should be able to incorporate their visions into the local authority planning processes and thereby achieve a consensus of visions. Local authorities have an important role because they can be either an enabler or disabler. For this reason WWF UK is focused on working with key groups to remove the potential and actual barriers facing communities who wish to engage in a more creative relationship with their own local authority. Nothing can be achieved, nor existing suspicions be exorcised, until all parties have built confidence and trust in each.

Since 1992/3 a growing number of communities, local authorities, NGOs and other voluntary organisations, and even the government (or at least the Department of the Environment pre-May 1997 – now the Department of Environment, Transport and Regions (DETR) have been generating a growing head of steam towards LA21. The Prime Minister is famously quoted for saying in New York, June 1997, that he wanted 'all local authorities to produce a Local Agenda 21 strategy by the year 2000'. Clare Short has refocused the Department for International Development (DIFD). Poverty alleviation is the aim, Africa is the priority, and improvements to water, health and education are the mechanisms. One can be excused for thinking that one can detect the traces of a commitment to sustainability! However, the MAI, which is being regarded as an economic development and wealth creation initiative, appears to be travelling in the opposite direction. It is an investor protection agreement that for the first time provides multi-nationals with new rights to protect their own interests at the expense of host communities and countries. The implications are profound but cannot be done justice here (see e.g. WWF briefing papers for more detailed information). As an illustration of what might be: local authorities would not be able to implement many policies to control inappropriate foreign investment, for example from highly polluting industries or ensure investment benefits to local communities through use of local supplies or personnel.

Under the similar North American Free Trade Agreement (NAFTA), Mexican local authorities are being sued over their refusal to allow US companies to establish toxic waste dumps. The US-based Ethyl Corporation is suing the Canadian government for $250 million over a ban on the dangerous toxin MMT (a petrol additive). Under MAI, governments would not have the right to restrict the investment of foreign investors. In Bermuda there has been a public campaign to stop McDonald's from establishing. Under MAI, McDonald's or anyone else would be able to set up in Bermuda as of right (examples provided in World Development Movement's *A Dangerous Leap in the Dark*, 1997). The MAI will restrict the powers of citizens to influence the pattern of economic development that takes place within their communities – one of the essential underpinnings of sustainability. In these situations it is not certain that even local projects such as LETS would not become a possible victim. The UK government has tabled 12 pages of amendments to the draft document, whereas the USA has tabled 600 pages. Why the discrepancy? The Department of Trade and Industry (DTI) who are leading the UK responses to this treaty appear to be unaware that MAI, as a closed club of the richest and friends of the richest counties who would be effectively protected by investment regulations, would perpetuate the poverty of the already poor and weakest economies of the world. One has to ask how DETR, DFID and DTI resolve their shared dichotomy – that is of course assuming they are aware there is a dichotomy!

Early in its work with local authorities, WWF UK identified and described a concrete block between the community at the bottom, and the local authority at the top. The characterisation described the concrete block from the view of those at the top, as a raft which conveniently protects them from too much interference from below. But for those at the bottom, the concrete block represents a lid which keeps them conveniently in place and quietened. The concrete is a real entity that can be recognised as an amalgam of the psychological, historical, economic, cultural and social inhibitors that have for so long effectively maintained the barriers between top and bottom, and therefore sustained a largely unchallenged status quo. The argument of the 'protect the status quo' lobby can be easily recognised – as environmental protection legislation will place us at a disadvantage with our competitors who do not have to meet the same standards, this will cost jobs. With this view reinforced by the MAI we might now place local, or even national governments at the bottom and the multi-nationals on top of a new concrete block!

A local sense

WWF UK is not a locally based organisation, it does not have hordes of people rushing around in wellingtons doing things for the environment. Where organisations such as the Groundwork Trust and British Trust for Conservation

Volunteers have locally based groups and projects, WWF UK's work at the local level is through project partnerships within the local infrastructure. As a result, during the period from 1992 to 1997 it established a number of partnerships with local authorities and small to medium enterprises (SMEs) in order to access local people and organisations. These partnerships have demonstrated on many occasions that the relationship between the local authority and their own communities leaves much to be desired and that many authorities still have to remove their concrete barriers before contemplating anything so risky as participative democracy.

There is growing use of the rhetoric of integrated strategic planning (ISP), and corporatism, which should mean that different parts of the same organisation all try to move in roughly the same direction. But it is with frustration that we have learned that although local authorities have many existing resources that could be used to deliver their ISP, too many fail to make the connection between these resources, their community development programmes, youth service, adult education or even community based social services and their LA21 community consultation programme. On the one hand, most local authorities already employ many people with the skills, expertise and local networks who are potentially so important for local implementation of LA21. Yet on the other hand they have a growing band of sustainability professionals who urgently need these very same skills and networks. Unfortunately, because expertise and needs are in different parts of the organisation, the two sides of the same authority are often unaware of each other. Memorably in one authority, on the very day it was planning its LA21 community based participative strategy, it made all its community educators redundant!

Building capacity

WWF UK's Local Agenda 21 strategy with local authorities and communities includes support to communities and local authorities in a two-way process to chip away at their concrete blocks. It is a strategy fundamentally committed to a philosophy of learning that is: empowering, participative and engaging of all parts of society; capable of developing a conscious awareness of rights and responsibilities through new insights on the world; able to promote the essential inter-relationship between social, economic and environmental components of sustainable living; underpinned by a local to global perspective that raises awareness of quality of life and equity issues; and supported by a long-standing track record of resources and materials developed and designed for specific audiences. WWF UK, often in partnership with external partners, has developed a range of LA21 products that are intended to support communities and local authorities in their new working relationship. However, as the work has progressed it has been realised that a growing number of strategically placed audiences, or stakeholders, needed to be supported. These include the often naturally occurring community champions; local authority elected members; local authority officers and senior managers; youth and people working with youth; teachers and teacher trainers; school managers; and SMEs. As a consequence, WWF UK has produced a wide range of new LA21 resources and training programmes that address the needs of these different stakeholders. The range of tools, such as consensus building, conflict resolution, facilitation, business planning, community indicators, and sustainability indicators, are tools not normally associated with the green movement. But LA21 has released into its new market place the opportunity to merge many established disciplines in a shared quest for sustainability.

The face-to-face approach of WWF UK's LA21 initiative has been a cascade model, where skills and experiences are developed at one level and then shared with another. The model begins mainly with those local authority officers responsible for implementing LA21 who in turn become part of the process of building the same skills and experience with community champions. The product of all this effort is often unpredictable but is aimed at achieving change in two main directions. First outwards from the authority towards the external community, and second inwards to other internal departments. It is reasonable to suggest that whilst many local authorities have become better at working with the external audiences, few have yet made the same progress internally. The corporatisation of sustainability across all functions of the local authority is still in its infancy, though there are some exceptions. Reading Borough Council for instance, has decided that the environment and sustainability will be the responsibility of all committees, not just an environment committee. Work through LA21 programmes must continue to produce Neighbourhood or Parish LA21 Action Plans, which in turn have to be seen as an essential part in the on-going process of informing local authority policy, improving community access to the decision-making processes and demonstrating the potential role of participatory democracy. Perhaps it is no accident that following the emphasis on community involvement by so many LA21 projects around the UK, that the government have initiated a national consultation called Modernising Local Government – Local Democracy and Community Leadership. This coincides with devolution in Scotland and Wales, but it will be some time yet before a judgement can be made on the rapid, even radical changes in the role and structure of regional and local government. But it is essential that the re-invention of local government does not dilute and distract the energy that has slowly begun to be spent on making local government more sustainable.

Beyond the local

WWF UK's Local Business Agenda 21 programme with NatWest Bank, created four SME thematic task groups to develop materials that would support the SME sector. The product of this relationship is the Better Business Pack which provides ready-to-use guides on transport, energy, waste and supply chain within the SME context. Because most businesses in the UK are SMEs, they are a major strategic target. However, the market research during the development of the Better Business Pack confirmed that the SME audience is a particularly tough nut to crack because as a sector there is little evidence of intrinsic motivation for working towards sustainability. Even the apparently obvious opportunity to improve profits is not a universal motivator. WWF (UK)'s work with this sector suggests that it is important to provide guidance on how to improve business practice by taking an environmental efficiency route rather than starting from an environmentally ethical position.

A guide commissioned by Forum for the Future and WWF UK in 1997, Involving Business in Local Agenda 21, reviews the extent to which business has become involved in Agenda 21. The guide concludes there is a 'patchy business involvement in Local Agenda 21', that there are 'many good individual initiatives' from which we can draw 'clear messages about how to spread good practice in the future' (CAG, 1998). But the questions are the same in the business sector as in the local government sector: Who is driving it? Is there the potential for increased critical mass, or will rivalry and commercial competitiveness undo the benefits? How do we get from the day-to-day operational imperatives to the higher level motivation that will enable the economic, social and environmental dimensions of sustainability to emerge? Chapter 30 (1: 525) of Agenda 21 begins 'Business and Industry, including transnational corporations, play a crucial role in the social and economic development of a country. A stable policy regime enables and encourages business and industry to operate responsibly and efficiently and to implement longer-term policies'. However, for this to be meaningful it is necessary for all three dimensions of sustainability to be able to work together towards new shared objects. Without a vision and the determination to implement it, competing agendas such as that manifesting itself through the MAI will continue to promote the interests of one group at the expense of others, and therefore undermine the long-term goals of sustainability. Therefore the continuing resolve to respond to the explicit LA21 agenda, together with the need to provide answers to the *between-the-lines* questions referred to earlier, must include at one extreme, the constant build up of local activity and experience that continually informs local policy development, and at the other extreme, a sophisticated working knowledge of the wider national and international perspectives that impact on economic development.

After development comes dissemination

WWF UK's future LA21 strategy is informed by the learning gained through four years of development and has two key elements. The first is called The Alliance for Sustainability, and the second is a two-stage programme of interactive seminars and in-house organisational appraisals, aimed at local government.

The Alliance for Sustainability is a partnership of local authorities and WWF UK. It intends eventually to expand to include other organisations of different backgrounds. It will work by identifying shared agenda issues that are suitable for collaborative responses, and develop these into a programme of work. To this extent the Alliance is about on-going development. However, it is also part of a dissemination strategy because its learning will be shared with other local authorities through publications, a web site and training programmes. There are several potential benefits from this new partnership, for instance: it will build on the relationship between a conservation NGO and local authorities, it will combine the learning of many disciplines, it will address real local issues in a practical and transferable way, it will produce new materials and validated research on a number of issues that are of social, economic and environmental significance. Most importantly the Alliance will provide a vehicle for the ongoing development of fresh and challenging thinking and action for sustainability.

There is perhaps a case to be made that there should be a moratorium on the creation of any new LA21 materials until what already exists has been fully utilised. There is a growing need to catalogue what is already available and to signpost people towards that which best meets their needs. Partly with this in mind WWF UK and the Local Government Management Board have jointly launched the two-stage Interactive Seminar and Organisational Appraisal programme. The full programme is intended to address the explicit needs of LA21, and to provide authorities with the information and guidance needed to develop a participative strategy for achieving its goals for sustainability. The programme will help local authorities to plot their course through the current plethora of LA21 materials and programmes by identifying the most appropriate for their needs. The programme promotes the notion of integration and holism by actively engaging elected members and officers of the authority alongside representatives of the community in a process which will encourage understanding of each other's perspective; of the strengths each can contribute; and a shared engagement in the development of new initiatives for sustainability. It is intended that policy and decision making will be better informed and take account of the needs of a wider constituency than is the tradition in most local authorities.

Participants in this programme have demonstrated varying degrees of confidence with the issues raised by LA21, and it has been interesting to note that even amongst the better-informed elected members, there is a degree of uncertainty of what is happening in their own area. Asked what was their main objective in attending, replies included: 'To make sure the County Council is actually implementing sustainable development' and 'To obtain a better idea of where the County is at the moment (not just the County Council)'. Others said they wanted to 'get a vision for action by members' and 'to broaden my understanding of Agenda 21 as environment spokesman on the County Council, and to sharpen my ideas of what can/needs to be done'. The same apparent uncertainty is to be found in some of the comments made by community representatives, who for example said, 'To understand what the County Council was doing in terms of Local Agenda 21 . . .'. Several community representatives seem to benefit from the contact with people from other agencies: 'The seminar gave me more contacts in organisations like the Police, who I may not have a chance to talk to, or even consider talking to'. Local authority officers seemed more comfortable with the concepts but more uncertain about the delivery and operational practice issues. Typically they said that they wanted 'wider understanding of process and practice', a 'clearer idea of the processes needed for a county wide Local Agenda 21 strategy', and information of 'key contacts with people in external organisations which will be useful in my area of work'. A glimpse of the role that LA21 can play in cross-linking came from a participant who was wearing at least two hats on the day: 'I am writing the Health Authority's Environmental Action Plan – I now have to re-draft and yes, this is a benefit!' (WWF UK and LGMB 1997–8)

The political scene

Agenda 21 is regarded by some as having failed, not so much because it has not yet brought about the actual changes that are needed, but more because it has so far failed to inspire the political leadership required to enable the necessary changes to be achieved. There are still wide gaps between the rhetoric and ambitions of LA21 and what is happening on the ground, where there are signs, although perhaps small, that the conceptual links are beginning to be made between environmental sustainability and the present unsustainability of human activity. There is some evidence that local authorities are seeing, but only slowly beginning to take action towards, sustainability as a core rather than peripheral function of local government. Reading Borough Council's shared responsibility for the environment across all committees is one example, and the City of Bradford Metropolitan District Council's, LA21 strategy which is led by the Director of Social Services is another. Most typically the LA21 initiative is housed in the Chief Executive, Planning or Environmental Services departments. In Bradford the move is towards a corporate approach where relationships with internal and external audiences and inter-departmental services are simultaneously working towards sustainability criteria.

At the local level there are new relationships growing between local government and the communities they represent because it is recognised that community involvement in LA21 is essential. Without the community's input, future commitment to sustainability policy risks being compromised because it lacks community ownership. LA21 is too radical and important to be left to traditional bureaucratic responses. Community ownership of both the process and product of LA21 is essential if new policies for sustainability are to be successfully implemented. Community involvement in a participative democracy paradigm, will make greater demands on political leaders, whose role, far from reducing, will grow in relevance. WWF UK has piloted LA21 work in rural parishes, where in one case, the LA21 process of community participation flushed to the surface and equipped a new breed of locally active democrats. 'I think we might have a real election for the Parish Council this year because so many new people are showing an interest' (Chairman of the Piddle Valley Parish Council 1997).

Half full or half empty?

The attitude taken in this chapter is essentially one of realistic optimism, though anyone who endured the New York Earth Summit 2, or read the subsequent press coverage, might be tempted to regard this attitude as an act of defiance in the face of reality! The lack of concrete commitments made at the Earth Summit 2 had little to do with Local Agenda 21 which was regarded as, so far, the most successful aspect of Agenda 21. Of course more could be done, responses to LA21 could be more ambitious and challenging, it should be funded and there is a case for saying that it should be a statutory rather than a voluntary requirement of local government. But any criticisms of the progress of LA21 should be accompanied with an honest assessment of the initial starting position at the time of the Rio Summit and the enormity of its task and ambition. So, is the glass half full or half empty? The Sustainability Indicators Report 1997, produced by West Devon Environment Network, and based on the responses of 1,300 local people, shows that 24 per cent of respondents had heard of Agenda 21, of whom only 5 per cent said LA21 had no relevance to them. Far from regarding this as a poor result, these figures suggest a remarkable awareness of issues which had been thought to be too abstract and removed from most people's reality. Surely the glass could be considered as half full?

In the five years between the Earth Summit and Earth Summit 2 in 1997, LA21 had an increasing impact in a growing number of areas, especially in local government. In reality, from its standing start in 1992 and up to 1997, there had been only three, or perhaps at best, four years of work, the rest being taken up with programme start-up time. Compare LA21 take-up and impact to that of equal opportunities and anti-racism legislation over the previous 20 years, and the response to LA21 begins to look good. In surveys carried out on behalf of the Local Government Management Board, (e.g. Tuxworth and Elwyn, 1996) more than 70 per cent of UK local authorities have committed to develop LA21 strategies. World wide, more than 1800 cities and authorities have made a similar commitment. It is important that the quality of engagement by these authorities should not be taken entirely at face value. Some LA21 strategies are no more than a list of things that need to be done, with little mention of the processes of engagement, impact on decision making, or links to the economic, social and environmental dimensions of sustainability. But a start has been made. In addition, there are countless small-scale community based environment and sustainability initiatives which also vary in their purposes. They range from uncomplex traditional environmental projects through to ambitious campaigns that have at their core an awareness of the complex inter-relationship between the environment and the negative forces that impact on it. There are many examples such as the Merseyside Wildflower Initiative, where neglected and unused land is turned into productive flower growing fields, simultaneously addressing income and employment generation, biodiversity, land reclamation and so on. The award-winning work done by the Coleraine Girls Secondary School in Northern Ireland, has for many years provided social interaction across the religious divide and other social groups through a land improvement project (examples taken from Warburton, 1995).

Much of the local level project work could probably have been achieved without Rio and LA21, but LA21 has provided a wider platform on which the importance of this work can be more dramatically demonstrated. A simple exercise which asks participants in a WWF UK Organisational Appraisal to brainstorm what is already going on in your area, in minutes, produced more than 50 named, and mainly voluntary sector initiatives alongside numerous local authority activities. But, just as Jonathon Porritt says of the business community in the foreword to 'Involving Business in LA21', 'it's becoming increasingly apparent that whilst each sector does its own commendable thing, they're not particularly good at combining forces' (CAG, 1998). How true also of local authority departments, and also between local authorities and their communities. One function of LA21 strategies should be to audit what is already happening, and identify potential opportunities for increasing the critical mass of activity. Only when that is done should new projects be designed to bring things together or fill the gaps.

So, in answer to the *between-the-lines* question asked at the outset of this chapter, perhaps the best we can say at present is that the chances of defending against the likes of the Multilateral Agreement on Investment are probably better than if the community and local government had not been energised by LA21. But only time will tell how successful we have been.

References

CAG Consultants (1998) 'Involving Business in Local Agenda 21– Guidance on Good Practice' unpublished document, WWF UK.

Chairman of Piddle Valley Parish Council (1997) Comments from an unpublished programme evaluation.

Desai, N. (1993) *Agenda 21 – Earth's Action Plan*, New York: Oceana Publications.

Hicks, D. and Slaughter, R. (1998) *Futures Education*, World Year of Education, London: Kogan Page.

Mayston, P. and Thornley. T. (1997) *Sustainability Indicators Report*, West Devon Environmental Network. C/o Bluebells, Brentor, Tavistock, Devon.

Robinson, N.A. (1993) *Agenda 21: Earths Action Plan*, New York: Oceana.

Tuxworth, B. and Elwyn, E. (1996) *Local Agenda 21 Survey*, LGMB: Environmental Resource and Information Centre, University of Westminster.

UN Department for Policy Co-ordination and sustainable development (1997) 'Programme for Further Implementation of Agenda 21', Advanced Unedited Text.

United Nations Commission on Environment and Development (1992) *United Nations Conference on Environment and Development Agenda 21*, Geneva: UNCED.

Warburton, D. (1995) *Starting with a Seed – a Guide to Community Projects in the Environment*, Godalming: WWF UK.

World Development Movement (1997) *Dangerous Leap in the Dark*, London: World Development Movement.

WWF UK (1997/8) 'Multilateral Agreement on Investment', various briefing papers, Godalming: WWF UK.

WWF UK/Local Government Management Board (1997/8) Interactive Seminar and Organisational Appraisal Evaluations, mimeo.

6 Children's participation in environmental decision making

Claire Freeman

Planning with children

Children constitute 23 per cent of the UK population, yet they are rarely considered and even more rarely consulted about the society in which they live (CSO, 1995, also see Knightsbridge-Randall, Chapter 7, this volume). They have for example no direct influence over the environment, public services or transport systems that they use, or the political agendas that affect them. There is no statutory representative mechanism available to children to influence and shape the society in which they are growing up and in which they will live their adult lives. This chapter explores how children's active participation in society can be encouraged so that they can contribute to shaping their future environment. It examines the potential role of Local Agenda 21 as one element in the process of encouraging children's participation.

Forces for change

The 1990s have been characterised by a growing awareness of the issue of children's participation in societal decision making in a variety of settings, and from a broad spectrum of organisations and agencies, ranging from national and local governments to local voluntary groups. The precise reasons for this growth are unclear, none the less certain influential trends are identifiable. First, the growing influence of the sustainable development movement, which implicitly places children at its centre with its concern for future generations – sustainable development is 'development which meets the needs of the present without compromising the ability of future generations to meet their own needs' (WCED, 1987). Hart in his seminal text, *Children's Participation*, further emphasises children's central role in sustainable development 'as tomorrow's adult citizens children must be able to create and manage sustainable communities' (Hart, 1997: 192). A second trend supporting participation has been the growth of work addressing children and the environment, primarily but not exclusively through the expansion of environmental education. Finally, there have been significant developments in promoting children's rights. The growth of interest in children's rights has not occurred in a vacuum but has been precipitated and supported by initiatives at both international and national level.

The Convention on the Rights of the Child

The UN Convention on the Rights of the Child has been critical in bringing to the fore the issue of children's rights in both global and national settings. The United Nations Children's Fund (UNICEF) described the Convention as follows:

> The Convention on the Rights of the Child, unanimously adopted by the United Nations general Assembly on 20 November 1989, is a legal document of unprecedented scope. With its 54 articles encompassing not just civil and political but also social and economic rights, the Convention stands alone as the single most comprehensive instrument of human rights law.
>
> (UNICEF, 1995: 1)

The Convention has been signed by over 177 countries, and was ratified by the UK in 1991 (Play-Train, 1995). Article 12 of the Convention enshrined the child's right to participate in decision making:

> States . . . shall assure to the child who is capable of forming his or her own views the right to express freely in all matters affecting the child, the views of the child being given due weight in accordance with the age and maturity of the child.
>
> (UNICEF, 1995: 5)

In the UK a national organisation called Article 12 run by and for children and young people, has been established to help children and young people find ways to make their voices heard.

Action Aid: Listening to Smaller Voices

Influenced by the Convention, an international initiative of significance is a project undertaken by Action Aid entitled 'Listening to Smaller Voices: Children in an Environment of Change' (Johnson *et al.*, 1995). The research focused on examining children's roles in the household, particularly their economic role in households in developing countries. Though the focus on employment is one that does not appear to offer immediate parallels to children's lives in the UK, the project's findings are both interesting and relevant. By virtue of the economic role that many children in developing countries have in their household, they are able in many cases to directly influence decision making in the household in a way that is not available to children in UK society. However, as identified in the postscript of the report, there are issues and problems concerning children's participation in decision making that are common to many children throughout the world, regardless of country of birth or societal status:

> The implications of misunderstanding the roles of children and their

> perspectives are that their quality of life may not improve and the future for their own development may be hindered rather than assisted. They are, after all, future mothers, fathers and decision-makers.
>
> (Johnson *et al.*, 1995: 86)

Agenda 21

Children's participation is an issue that transcends national boundaries and which requires concerted efforts at global, national and local scale. Agenda 21 clearly recognises the significance of children's participation in promoting both the global and local environmental agenda:

> The involvement of today's youth in the environment and development decision making and in the implementation of programmes is critical to the long term success of Agenda 21. . . . It is imperative that youth from all parts of the world participate actively in all relevant levels of decision making because it affects their lives today and has implications for their futures.
>
> (LGMB, 1992: 224)

It goes on to exhort governments to 'establish a process to promote dialogue between the youth community and Government at all levels. . . . Establish procedures allowing for consultation and possible participation of youth. . . . Promote dialogue with youth organisations' (LGMB, 1992: 224).

As one of the key agreements at Rio, Agenda 21 has international and subsequently national support, with children as a significant element. However, as events at Rio indicate, even where there is positive support for children's participation children can still experience difficulties in getting themselves heard. At Rio the official youth delegates were promised one hour during the 14-day Summit to speak. In reality they had only ten minutes, and when they tried to speak to the press they were arrested by the UN police for holding an 'illegal press conference' (Peace Child International, 1994: 80).

In 1994 a children's edition of Agenda 21 entitled *Rescue Mission Planet Earth* was published by Peace Child International. The book represented the efforts of some 10,000 children from about 100 countries and is clear testimony to children's concern for their environment and their future, and their ability to contribute positively to shaping this future. In the book the children call for 'a global democracy of children' and grapple with the problem of 'how can 2.5 billion human beings under the age of 18 be connected in a way that would be democratic without being bureaucratic?' (Peace Child International, 1994: 84). A pre-requisite to any degree of global democracy is national democracy and local democracy. In the UK, although positive steps are being made towards developing children's participation, it is debatable whether these constitute any significant move towards children's democratic empowerment.

Seen and not heard! The UK scenario

Children remain largely unheard within the UK as a whole when it comes to shaping society and determining the future environment in which they live. The UK has never had a good record on children's rights and lacks an overtly child-friendly culture. Children have no representative at national level in decision making. There appear to be no national governmental initiatives that focus on children's participation in decision making, and there is no national guidance on children and Agenda 21. However, in March 1995 the Children's Rights Office opened in the UK, continuing the work of the Children's Rights Development Unit. The Office is a registered charity and receives no government funding. The purpose of the Office is two-fold: to encourage implementation of the UN Convention, and to promote the case for a Children's Rights Commissioner to address the issue of children's lack of voice, describing children as '13,000,000 citizens with no voice in government'. It is as yet to early too gauge the effectiveness of the Office in addressing the issue of children's rights to participate in decision making at national level, though the Office has already made significant progress in raising general awareness of children's rights issues.

Whilst children's voices remain largely silent at national level in decision making, there is welcome evidence that at local authority level there are positive indications that children's issues are increasingly being seen as legitimate local authority concerns, as indicated by the growing number of local authorities signing up to the UN Convention. By May 1997 the Children's Rights Office recorded a total of 367 organisations having signed up to the UN Convention, including 78 local authorities. In addition to developments at the wider local authority level there is also evidence of a growing recognition by local authority departments such as planning (e.g. Freeman, 1995; Hart, 1997; Smith, 1995) that children have a right to be consulted. How to achieve participation however presents a challenge that few individual departments and few local authorities as a whole seem as yet to be capable of meeting.

Children's participation: the challenge

In addressing children's participation there are certain fundamental considerations that any local authority or organisation will have to take into account in order to address the practicalities of promoting children's participation.

How much participation?

Once a commitment has been made to children's participation, a critical appraisal has to be made of what is meant by participation, how much participation and how much power in decision making will be devolved to children. It is essential that the reasons why children's participation is being encouraged and the expectations of all participants are clear. For many, particularly those in local authority departments such as planning, participation is a radical departure from traditional methods of working, it is not easy and there are pitfalls. Hart (1992) in his study of children's participation for UNICEF, identified a number of levels at which children can participate, ranging from tokenistic to child-centred. Hart adapted Shelley Arnstein's ladder of community participation to create a children's ladder of participation (see Figure 6.1). To be meaningful participation must be on the higher rungs of the ladder and not at the lower tokenistic levels.

1. **Child initiated, shared decisions with adults:** children have the ideas, set up the project and come to adults for advice, discussion and support. The adults do not direct but offer their expertise for children to consider.
2. **Child initiated and directed:** children have the initial idea and decide how the project is to be carried out. Adults are available but do not take charge.
3. **Adult initiated, shared decisions with children:** adults have the initial idea but children are involved in every step of the planning and implementation. Not only are their views considered, but they are also involved in taking the decision.
4. **Consulted and informed:** the project is designed and run by adults but children are consulted. They have a full understanding of the process and their opinions are taken seriously.
5. **Assigned but informed:** adults decide on the project and children volunteer for it. The children understand the project and know who decided why they should be involved and why. Adults respect their views.
6. **Tokenism:** children are asked to say what they think about an issue but have little or no choice about the way they express those views or the scope of the ideas they can express.
7. **Decoration:** children take part in an event, e.g. by singing, dancing and wearing T-shirts with logos on, but they do not really understand the issues.
8. **Manipulation:** children do or say what adults suggest they do, but have no real understanding of the issues; or children are asked what they think, adults use some of their ideas but do not tell them what influence they have had on the final decision.

Figure 6.1 The children's ladder of participation (after Hart 1992)

Whose responsibility?

If children are to become effective participants in shaping the local environment then certain fundamental pre-conditions are required. The professions need to acknowledge that they bear some responsibility for the situations where children are excluded from decision making and conversely that as professionals they have a responsible role to play in reversing this exclusion. On the positive side however, there is an increasing body of evidence which suggests that there is a growing consensus among a wide range of professions, agencies and organisations that children's rights to be consulted and participate in decision making are a legitimate and critical concern. Mechanisms will need to be put in place to facilitate exchanges of information, and expertise between all those who work with children and who can contribute either directly or indirectly to providing for children's participation in environmental decision making.

A role for LA21

Agenda 21 is important for local authorities and organisations as it offers a way forward and its endorsement by national government provides positive support for local authorities and organisations in producing their Agenda 21. It represents a radically different approach to local environmental planning and decision making in that it emphasises the need for co-operation, making partnerships and integrative working at all levels. It promotes ways of working that cross-cut traditional professional and organisational barriers. To effect children's involvement in shaping the environment, support is required from all sectors of the professional and academic community and a recognition that the traditional pattern of confining children's issues to fields such as education and health is no longer adequate. LA21 incorporates an immense breadth of subject matter, none of which is the exclusive preserve of any one discipline, organisation or department. To be effective a child-centred LA21 must of necessity be diverse in content and structure and be inclusive, acting as an umbrella bringing together and supporting existing and developing participatory initiatives. The next section explores some child-centred participatory initiatives which organisations and local authorities have developed and which can usefully contribute to a child-centred LA21 process.

Developing participatory initiatives

A range of participation initiatives being developed for children and young people is apparent throughout the country (Freeman, 1996). These initiatives are often hampered by a lack of overall co-ordination and exchange of information nationally and even locally. Frequently, groups and organisations in the same locality do not know what others are doing. As yet there is no fully developed functioning national programme for children's participation evident in the UK and no local authority with a fully developed children's LA21 participation programme. Several cities are however, promoting and working on developing a children's agenda, as with the child-friendly city programmes being developed by Edinburgh, Leicester and Manchester. Leicester also has a well-developed Agenda 21 programme 'Blueprint for Leicester'. When Manchester began developing its LA21 in 1993, it was very concerned that it should not just be another bland local authority document, and that children's voices should really be heard. It appointed a LA21 Children and Young People's officer for this purpose. Manchester's concern for including children as active participants has since been mirrored in a number of other local authorities which have adopted a range of methods to facilitate children's participation, including:

- children's councils
- children and young people's forums

- newsletters, circulars and leaflets
- questionnaires and consultation exercises
- participation through drama, games and environmental events
- school-based projects
- establishing information exchange facilities
- setting up professional worker liaisons and networks
- support workers including children's LA21 officers
- conferences
- partnerships between children's organisations and local authorities.

Selected examples are examined here in the context of the Leeds local authority. As a local authority Leeds has committed itself to promoting children's participation in shaping the future of the city and has developed a range of initiatives to enable it to move towards this goal.

Participation in LA21: A case study of Leeds

In recent years Leeds has directed serious attention towards including children, not only in its LA21 process but more significantly as part of their wider on-going process of developing public participation in decision making. There are however, key issues that need to be considered and addressed if effective participation is to be achieved.

Developing an informed base

Two major consultative exercises have been undertaken in the city. The first, Leeds Listens (Burden and Percy-Smith, 1996), was a joint exercise between researchers at Leeds Metropolitan University and the city council. At the time it was undertaken it was the largest and most comprehensive survey of its type undertaken in any UK city. Over 2000 children and young people were consulted using a range of methods including questionnaire surveys (for the over-10 age groups) and the establishment of a number of focus and discussion groups (for all ages from pre- to post-school age). The rationale for the exercise was to build on the 1995 audit of children's services in the city 'Seen, Heard and Listened to', in order to understand 'more about how children and young people see their city and to explore their hopes' (Burden and Percy-Smith, 1996: iii). In the section entitled 'Children and Young People Today', Cheeseman reflects on the value of research as part of the participative process, as follows:

> Studies of childhood and children demand a breadth of vision which should illuminate the key areas that constitute children's lives. . . . The purpose of reports such as this must be not only to mirror what is happening now, but to reflect on the past and gain insights which enable us to influence the future.
>
> (Burden and Percy-Smith, 1996: 9)

It is still too early to assess the value of Leeds Listens research in terms of its impact and results. The research is intended to inform local authority policy and to draw up an action plan which ensures that children are at the centre of future planning in the city.

A second research study, funded by the Gulbenkian Foundation, was carried out as a joint venture between the Leeds Environment City Initiative and the northern office of the Community Development Foundation, with a consultation draft released in November 1996. The aim of the study was to establish how children and young people could be involved in decision making in Leeds, and identify the types of participation processes which would be most appropriate. A parallel concern was to identify how Agenda 21 can play a role in this process. As in the Leeds Listens research, a range of children, young people and interested professionals were consulted. Interestingly, the research found that though environmental concerns have a high priority for young people, 48 per cent felt powerless to do anything as no one would listen to their ideas. It was a finding that clearly demonstrated the need for the city to address the issue of children's participation.

When asked about which methods of participation should be developed, the idea of a city-wide, single-issue forum was not received with much enthusiasm by the children and young people consulted. Participation methods which were seen to be more attractive were those that focused on working with children and young people in groups across the city. Locally accessible facilities which provide for a range of services and which are not just concerned with environmental issues were seen as the most promising way forward by the young people. Positive support was also expressed by young people for establishing adult advocates to work with and facilitate participation and decision making on issues ranging from mainstream environmental concerns to wider and often more significant concerns such as crime and unemployment.

The findings of both research exercises clearly indicate that despite their clearly expressed wish to participate in shaping their local environment, children and young people feel alienated from existing decision-making processes and limited in their ability to contribute. It also became evident that there is no single preferred universal participative mechanism, and that effective participation demands the use of a diverse range of methods to meet the needs of a diverse range of participants.

Working together

Partnership is one of the key ideas of the 1990s. Partnerships are particularly crucial in developing children's participation, as knowledge, expertise, experience and resources are rarely the preserve of any one organisation or local authority department. Leeds has been home to a number of partnership initiatives including the Children's Participation Project developed by Save the Children and used in a number of projects elsewhere in the country. Liaisons and partnerships are important to Save the Children, which as a charitable organisation is unable to fund and resource projects alone. In partnerships it is able to offer its considerable expertise in working with children. Two partnership projects which it has developed in recent years include: the Children's Participation Pack, developed in partnership with Kirklees Metropolitan Council (Save the Children, 1996) and the Children's Rights Service, developed in partnership with Leeds Social Services. The Service is an independent support service based at Save the Children for use by children and young people in care.

In addition to partnerships there are also networks, which tend to be less formal associations focusing on exchanges of information and experience. Networks in various stages of development in Leeds include the Leeds Play Network, which in addition to addressing children's play also works on encouraging participative activities particularly through its contacts with Playworkers. The network has expressed strong support for the appointment of a funded adult worker whose core task would be to develop and support child-led initiatives across the city. A Voluntary Sector Youth Forum, part funded by Community Benefits and Rights in the city council, was established in 1997 with the aim of promoting professional voluntary sector services for young people. Whilst the focus is on networking between adults actively working with children, the intention is to promote the involvement of young people in the planning and provision of services, providing training as appropriate for the young people involved.

Child and youth-led participative fora

Since the late 1980s there has been support in the local authority for developing Youth Fora. Youth Fora and Councils have been used successfully in promoting children's participation, at both international and national levels. France has some 740 children and youth councils, which have the right to make suggestions to the adult councils. Belgium, Switzerland, Austria, Germany, Poland, Hungary and Italy have also developed similar councils. Such councils are rare in the UK but there has been some progress by local authorities – Devon County Council for example has actively promoted the development of a Youth Council, Manchester has a Young People's LA21 Forum and Nottingham a Youth Environment Forum. Leeds has a small number of Youth Fora in the city which are locally based and generally focus on specific groups of young people – such as the Black Youth Forum based in Chapeltown and Azad Youth Forum for Bangladeshi, Sikh and Pakistani young people in South Leeds.

Whilst the Leeds Listens and Gulbenkian/Leeds Environment City Initiative revealed that a city-wide Youth Forum would have limited appeal for and support from children and young people, the City Council is still keen on developing Youth Fora as a means of giving a voice to younger residents in Leeds. Consequently, over the next two years the council is to set up eight fora across the city. The challenge will be to ensure that these fora, particularly in the early stages are successful in encouraging the participation of children and young people who are known to experience particularly significant levels of alienation and deprivation.

Resources and support

A pre-requisite for all successful types of community participation is support, for as Checkoway *et al.* stated, 'people cannot be expected to participate effectively if they lack knowledge, skills and attitudes conducive to the task' (1995: 137). In his *Guide to Effective Participation* Wilcox (1994) similarly explores the issue of support in community planning, and identifies capacity building as an integral component of such support. He defines capacity building as 'training and other methods to develop the confidence and skills necessary to help people to develop the confidence and skills necessary for them to achieve their purpose' (p. 31). For children and young people such capacity building is critical if participation is to work.

A recurrent issue in developing children's participatory initiatives is that the resources necessary to support participation are under-estimated by resource providers. Maggie Goodman, the Article 12 young people's network adult support worker, in common with many other support workers working both nationally and in Leeds, stresses the fact that the success of initiatives such as Article 12 demands extensive levels of trained and resourced adult support. The support necessary takes two forms: support directed at the participating children and young people and support and possibly training for the adults involved. In developing LA21 initiatives, support for the adults involved may be especially important as many of those involved in the LA21 process are planners, architects, housing managers and others who generally have little experience of working with children and young people. In developing a child-centred LA21 it will be necessary therefore to ensure that resources are available to support a range of participative projects, not all of which will fall directly under the LA21 umbrella.

Bringing it together

In order to be effective it is essential that projects encompass a diversity of participation methods and situations, and demonstrate the flexibility to adapt to local needs and circumstances. One project that achieves both of these is the Cross Flatts Park project in Leeds, which brings together a diverse range of participants and uses a variety of participative techniques.

Reclaiming the park

Children at Cross Flatts Park Primary School undertook a project during 1996 that contains many of the elements required for a successful participative LA21-type project (McGeever, 1996). The school where the project was based is situated in Beeston, an ethnically mixed, inner-city neighbourhood dominated by terraced housing, where green space is at a premium. The project was a partnership between Health for All, Cross Flatts Primary School, South Leeds Team Ministry, Groundwork, the pre-School Learning Alliance and local residents, with further assistance from a student at Leeds Metropolitan University and British Telecom's Community Action Award Scheme. Beeston Health Group and local children identified dog fouling and poor and vandalised facilities as significant factors preventing them from making full use of the local park. A project using photos and Planning for Real techniques was undertaken where pupils from the school created models of the park to show the existing conditions and what the park could look like if improved. A number of methods were used by the children to draw community and city-wide attention to the Cross Flatts Park project, including a local questionnaire survey, placing yellow flags in the park close to dog fouling to show the problem, a public launch to which the local Member of Parliament, councillors, city council officers, and the local press were invited (invitations were also sent to the Queen and political leaders).

In the short term the project has proved to be a very effective participation project. In the longer term the success of the project will depend on the response of the city council and others to the issues raised by the children, notably dog fouling and poor facilities in the park. If no improvements are made and the poor conditions in the park are not addressed by the local council it could signal to the children that the issues they identify as important and the effort they put into the project are not reciprocated. In such a situation the danger is that the children would feel alienated from local decision making and be reluctant to take part in any future projects. In short, if participation has no results, then it has not been effective, and for children it is particularly important that results are rapid and tangible.

Developing children's participation

For local authorities and organisations there are therefore a number of issues that can be identified as critical if successful participation initiatives, located on the higher rungs of Hart's 'Ladder of children's participation' are to be realised. In the absence of national guidance and proven participatory methodologies for involving children and young people in local environmental planning and decision making, organisations and local authorities will need to develop their own participatory process. LA21 offers a great opportunity for all those interested in promoting the participation of children, enabling children and young people to shape their future society. It provides a new and exciting way of bringing professions and organisations together. If local authorities and other organisations are to ensure that LA21 is participative and effective in encouraging children's participation, it will be necessary to:

- *Establish clear aims and vision*: any organisation seeking to work with or enhance its work with children will need to have a clear understanding of why it seeks to involve children and what it hopes to achieve through such involvement.
- *Be informed*: before any participative process can be initiated it is essential that there is an understanding of what it is that children need and want. It may be necessary to undertake a preliminary audit.
- *Be action focused*: if children's participation is to be encouraged and interest sustained it will be necessary for LA21 to focus on and encourage

participation in issues and projects which children feel that they can directly influence and where the results of their efforts are tangible.

- *Recognise diversity and provide flexibility*: it is necessary to recognise both the diversity of children's lives and needs, and to provide different ways in which children can participate and at different levels.
- *Use environment in its broadest sense*: LA21 is not just about environmental education, where green issues at the forefront tend to be school grounds, recycling, tree planting and litter. Environment in LA21 should incorporate the breadth of environmental issues which could include issues such as poverty, crime and bullying.
- *Promote co-ordination*: LA21 necessitates sharing experience within and between professions and organisations, and working together as appropriate.
- *Offer real commitment*: recognising that such commitment will need to be supported by the provision of long-term resources. At present many LA21 initiatives are dependent on short-term funding, and many LA21 support workers are employed on a short-term basis.

What of the future?

At this stage no local authority in the UK has a fully developed Children's LA21 programme, developments are fragmented, thin on the ground and in most cases in their infancy. It is up to local authorities and organisations working with children to embrace the immense opportunities offered by LA21 as a means of enabling the UK's 13,000,000 voiceless citizens (Children's Rights Office, 1995) to determine their own environmental future. Although progress is being made in cities such as Leeds, overall progress is slow and superficial and much more needs to be done if children and young people are really to become active citizens in the twenty-first century. LA21 requires all those with a concern for children and young people to come together, and share their expertise and knowledge. Only by sharing can real progress be made towards developing a better future for and with our children and young people.

References

Burden, T. and Percy-Smith, J. (1996) *Leeds Listens to Children and Younger People*, Policy Research Institute, Leeds: Leeds Metropolitan University.

Central Statistical Office (1995) *Annual Abstract of Statistics*, London: HMSO.

Checkoway, B., Kameshwari, P. and Finn, J. (1995) 'Youth participation in community planning: what are the benefits?', *Journal of Planning and Educational Research*, 14(2): 134–139.

Children's Rights Office (1995) '13,000,000 Citizens With no Voice in Government', London: Children's Rights Office.

Freeman, C. (1995) 'Planning and play: creating greener environments', *Children's Environments*, 12(3): 381–387.

Freeman, C. (1996) 'Local Agenda 21 as a vehicle for encouraging children's participation in environmental planning', *Local Government Policy Making*, 23(1): 43–51.

Hart, R. A. (1992) *Children's Participation from Tokenism to Citizenship*, Florence: UNICEF.

Hart, R. A. (1997) *Children's Participation: the Theory and Practice of Involving Young Citizens in Community Development and Environmental Care*, London: Earthscan.

Johnson, V., Hill, J. and Ivan-Smith, E. (1995) *Listening to Smaller Voices: Children in an Environment of Change*, Somerset: Action Aid.

Local Government Management Board (1992) *Local Agenda 21, Supplement No. 2. Agenda 21 A guide for local authorities in the UK*, Luton: LGMB.

McGeever, P. (1996) *Reclaiming the Park: a Project to Improve Cross Flatts Park*, Leeds: Health for All.

Peace Child International (1994) *Rescue Mission Planet Earth. A Children's Edition of Agenda 21*, London: Kingfisher Books.

Play-Train (1995) *UN Article 31 Pack, Children's Rights and Children's Play*, London: National Play and Information Centre.

Save the Children (1996) *Children's Participation Pack*, London: Save the Children.

Smith, F. (1995) 'Let the children say', *Town and Country Planning*, September: 230–231.

World Commission on Environment and Development (1987) *Our Common Future*, Oxford: Oxford University Press.

7 Youth and Agenda 21

Jane Knightsbridge-Randall

> Young people are more widely travelled and educated than any previous generation. They are not parochial or insular; unlike older folk. The message for newspapers is simple: don't believe your own prejudices. Youth will not let you down if you do not let them down.
>
> (*Guardian*, March 1997[1])

Public perception of youth is often shaped by the media, where young people are either characterised in terms of anti-social behaviour or portrayed as '15–24 year olds with an addiction to cool culture, computers, consumerism, clubs and an aversion to anything that demands more than a three minute attention span'.[2]

Over the past ten years there have been marked changes in young peoples' situation. Eighty per cent of young people go on to full-time further education and training. With the virtual disappearance of the youth labour market, the decline in young people's wages and unwilling economic dependency on parents or state, we are experiencing a steady extension of the period of youth. The boundaries between childhood, youth and adulthood have become increasingly blurred. Transitions of youth from childhood to adulthood have been regarded as being akin to a rather vicious version of the board game snakes and ladders (Coles, 1995). Statuses are constantly in flux (Jones and Wallace, 1992) and subject to complex processes of negotiation and renegotiations 'between young people and their families, their peers and the institutions of the wider community' (Jones and Wallace, 1992: 4). 'These are the key actors in the social construction and reconstruction of the statuses of youth' (Coles, 1995: 4).

Protest against road building has produced a new youth phenomenon. Young people who get involved and take action, often find themselves featured in the media first as part of an alternative counter-culture, then a picture unfurls, framing the young person as a member of an ordinary family. In the 1980s and 1990s negative representations in the media have foreshadowed policies written in reaction to a moral panic: examples include the Criminal Justice and Public Order Act 1994, creating the offence of aggravated trespass which has implications for those opposing fox hunting, involved in roads protests and attending free festivals. The Social Security Bill, secure training centres, housing legislation for single mothers, welfare to work and education fees are key themes in current and proposed legislation and have significant implications for the quality of young peoples' lives.

There has never been a generation more informed about environmental issues than this one. For youth in the 1990s, some of the issues they have studied in school are publicised on world wide TV culture and interpreted locally in ways specific to each culture and its setting. What this means is that we have a very well-informed youth, armed with the knowledge and argument to take on these issues if they so desire. With media exposure and the inclusion of social and environmental issues as part of the school curriculum, no issue is taboo.

Young people know that problems cannot be tackled overnight and that a sustained effort is required. They want to be involved and they want their voice to be heard. The British Youth Council, in a briefing paper for politicians and key decision-makers, points to young people's growing concern with so-called 'single issues' such as the environment, human rights, racism, the Third World and animal rights. They say that 'it is quite clear that pressure groups often provide an environment that makes young people feel that their opinions matter, their involvement is valued and, crucially that they as individuals can make a difference' (British Youth Council, 1995).

> In everyday life, we function within our environment and just as we affect it, it affects us. Our environment has physical, psychological, and social aspects which profoundly influence our behaviour and experience. We are constantly interacting with our world and are inseparable from it; we can have no meaningful sense of ourselves without a heightened awareness of the world in which we live.
>
> (Benedetti, 1990: 77)

All young people have needs that relate directly to the environment in which they live. Agenda 21 was agreed by all 179 nations present at the United Nations Conference on Environment and Development (UNCED) or Earth Summit in Rio in June 1992. It addressed the critical issues of today, and aimed to prepare the world for the challenges of the next century. It is intended to be dynamic – evolving according to the needs and circumstances of the various countries. The spirit of the Summit was captured in the 27 principles of the Rio Declaration. Principle 21 states: '*The creativity, ideals and courage of the youth of the world should be mobilised to forge a global partnership in order to achieve sustainable development and ensure a better future for all*'.

Agenda 21 promotes participative democratic processes, which enable people to contribute at all levels towards the development of policies that are socially, economically and environmentally sustainable. Agenda 21 recognises the critical importance of involving young people by 'advancing the role of youth and actively involving them in the protection of the environment and the promotion of economic and social development' (Agenda 21, Chapter 25). The preamble to the section on strengthening roles asserts that critical to the effective implementation will be the commitment and genuine involvement of all social groups.

However, there are crucial issues facing the discussion of the efficacy of Agenda 21, not just concerned with the nature of the actual document itself, but arising from its relationship with the wider social order, in all its discursive and institutional complexity. Texts carry with them both possibilities and constraints, contradictions and spaces. The reality of implementing Agenda 21 depends upon the compromises and accommodations of these textual possibilities, constraints, contradictions and spaces. Environmental issues are seeping into the political agenda of most governments and many international agencies; the process is one of action and reaction.

Barthes in *S/Z* (1975), made the distinction between two kinds of text, that which is lisible (readable), and that which is scriptible (writable). The translator, Richard Miller uses the terms 'readerly' and 'writerly'.

> The readerly text is based on logical and temporal order, it communicates along a continuous line, we read it one word after another, we consume it, passively. . . . Writerly texts make us not consumers but producers, because we write ourselves into it, we construct meanings for it as we read it, and ideally these meanings are infinitely plural.
>
> (Lodge, 1977: 66)

Elam (1991) also made a distinction between two kinds of text: he identified these as the theatrical (or performance) text and the written (or dramatic) text. Within the world of theatre, the written text is determined by its very need for stage contextualisation. 'As Paola Gulli Pugliatti (1976: 18) has said "the dramatic text's units of articulation should not be seen as units of the linguistic text translatable into stage practice", but rather as a linguistic transcription of a stage potentiality which is the motive force of the written text' (Elam, 1991: 209).

What this suggests is that the written text/performance is not one of simple priority but a complex of reciprocal constraints constituting a powerful inter-textuality. Each text bears the others' traces. Agenda 21 was not established in abstract and applied in abstract. It is concerned with the inter-textuality between written text and performance text, which is encoded, decoded and transcodified.

This leads us to approach Agenda 21 and Local Agenda 21 as a discourse, with 'actors modulating between their own sense of the text and the shape the director is giving, the actors then find ways to make that shape their own' (Harrop, 1992: 63). Writing on the failure of educational reform, Sarason says that we have still not learned to focus our efforts on understanding and working with the culture of local systems:

> Ideas whose time has come are no guarantee that we know how to capitalise on the opportunities, because the process of implementation requires that you understand well the settings in which these ideas have to take root. And that understanding is frequently faulty and incomplete. Good intentions

> married to good ideas are necessary but not sufficient for action consistent with them.
>
> (Sarason, 1990: 6)

Chapter 28 of Agenda 21 states 'All local authorities in each country should be encouraged to implement and monitor programmes which aim at ensuring that women and youth are represented in decision-making, planning and implementation processes'.

A number of international instruments call for increased consideration of the perspective of youth in local decision making. The UN Convention on the Rights of the Child (1993) enshrines the right of all young people to civil liberties, including participation in decision making, freedom of expression, freedom of association, freedom from discrimination and freedom for all forms of violence. (This Convention is further explored in Freeman, Chapter 6, this volume.) Agenda 21 identifies youth as one of the major social groups who have a stake in health, economic development, housing, transport and racism issues, and calls for their consultation in the development of Local Agendas 21.

The reality is that up until the age of 18 children and young people have no vote and can play no active part in the political process, so children and young people remain a low priority and issues are seldom considered from the perspective of children and young people.

Review of youth/young people's involvement in LA21

Research undertaken for the WWF (UK) (Knightsbridge-Randall, 1996) has shown that most LA21 development plans in the UK have not incorporated contributions from young people. Most local authorities seem to doubt the value of youth contributions. Even those that are willing to consider the value are uncertain how to draw in effective youth involvement. The research also showed, however, that where youth have been engaged, they added value to the development of LA21.

Over 200 contacts were made, including local authorities, voluntary groups, special interest groups, youth workers, environmental co-ordinators, environmental organisations and the Department of the Environment. Information was obtained from organisations in England, Wales, Scotland and Northern Ireland; both rural and urban perspectives were represented. Conscious of the observation that environmental issues are often marked by a gender, social class and ethnicity bias, the report highlighted the vital role of under-represented groups. Authorities were identified and reached in various ways. Some were contacted because of the high profile they have in LA21. Others were identified in a research project for the International Community Education Association (Kightsbridge-Randall, 1995). Personal knowledge and connections produced other sources. Information was provided by the Local Government Management Board and the London Ecology Centre; other contacts came via WWF UK and individual contacts were made through networking.

The awareness level of Agenda 21 was very low. Even in authorities with local Agenda 21 co-ordinators there was a lack of knowledge among people not directly involved in the initiative. The following comments are examples of the responses received: 'Unfortunately I have no idea about Agenda 21, but thankfully we do know about good practice with young people in the environment'; 'We don't know anything about Agenda 21 I'm afraid. . . . Is it important? . . . perhaps it doesn't affect us in Scotland' (Knightsbridge-Randall, 1996).

A number of interviewees expressed disappointment that their attempts to involve youth/young people had not been successful. Some were keen and anxious to develop work which was relevant, but were tied to structures already established that were not accessible to young people. Others did not recognise this, and young people were accused of not showing an interest. Blame was also levied at other adults for not nominating a young person for one committee or another. There were authorities that had not considered youth in their plans for LA21.

All 33 London authorities were contacted by letter; there were 17 responses. One person wrote that the involvement of young people was 'a bit of a hole'. Two authorities had organised conferences and others were planning to have: a conference/a children's summit/a youth conference. School newsletters were produced by several local authorities and two authorities referred to eco-school schemes and one had a video recording of a school's environmental play. An environmental co-ordinator said they were going to try and persuade the authority to appoint a youth worker with responsibility for Agenda 21. Another co-ordinator said that because of legal difficulties children under the age of 18 could not be involved in forums. A London Agenda 21 initiative had been established to address the sustainability of the capital; 11 stakeholder groups were involved in examining and making recommendations on specific London-wide issues. London's young people were represented by executives from the National Union of Students.

In an international review of the London Agenda 21 plan, Lawrence wrote:

> If the London metropolitan area is to become more sustainable it will be because the people have chosen that path, not because of the plan. The plan will describe a journey and a destination. Individuals will need to decide the trip is worth making. What will inspire them to do so?
>
> (Lawrence, 1996: 7)

If sustainability is dependent on the choices that people make then where are young people in this debate? And what opportunities are there for them to be involved as active participants in the decision-making process?

Cultural differences

Although children and youth are identified for strengthening roles in Agenda 21 young people have been missing in most local government's development of

LA21. When involving them at a late stage there has been a tendency to try and graft them on to existing LA21 structures and strategies, which usually take the form of roundtables and focus groups. There is also an inherent problem with these approaches as they tend to fragment the issues and not reveal the connections. Then young people are expected to make the very connections that decision-makers have constructed. The WWF UK research (1996) recommended that the most effective way of involving young people is to start from young people's own enthusiasm rather than established practice.

There were some interesting environmental initiatives taking place. These were operating on tiny budgets, relying on the dedicated commitment and enthusiasm of participants to take them forward. These peripheral youth initiatives were marked by continual struggle but the message and lessons on how to involve young people came through loud and clear. The following are examples from the WWF report (1996). At a sixth-form conference in the Yorkshire Dales young people voiced their thoughts on the future. One stated 'environmental care should definitely not be left to the politicians; each and every one of us must strive for a better environment'. In Oxfordshire the themes that the group wanted to cover were housing; work; resource management; agriculture; biodiversity and transport. Their aim was to identify the issues and encourage thought regarding viable solutions. Brighton's young person's group VOYCE (Views of Young Concerned Environmentalists), for young people aged 11–25, met once a month to identify issues and take action. In Manchester it was learnt from experience that the traditional adult approach to forums did not work and that a variety of approaches and creative methods have to be used.

When Devon Council decided to follow the Canadian example of consultation by establishing roundtables for LA21, it was clear to the youth council that a roundtable used as a forum for discussion by adults would not be appropriate for young people. The youth council wanted to make the most of the opportunity for young people to express their views about the environmental factors which greatly affect the future of our planet and society. Following research the young people were so frightened of what they discovered, they resolved that the only way to make real change would be to utterly revise their day-to-day behaviour. The nation's concern with VE day in 1995 caused them to reflect on who or what was the common enemy of the late 1990s and they concluded that the common enemy is potential environmental disaster. The youth council identified, through consultation with young people, that using the arts as a medium for education was extremely effective. It also noted that complex ideas were often much better understood by young people when directed through the arts rather than the traditional lecture theatre. The Devon project involved young people expressing themselves honestly and in any way they chose, whether through language, drawing, music, dance, drama or something completely different.

Natural Allies was a collaboration between Queens Hall Arts Centre, Northumberland National Park, Hexham Park School and Priory School (both special needs schools) in the autumn of 1995. The approach to raising awareness of both the environment and young peoples' needs was innovative and challenging. It was an eight-week residential project that set out to challenge and extend the participants' needs, and in doing so to improve 'quality of life'. The group wrote, 'bringing arts and environment together in the project will enable us to facilitate an experience which has lasting effects and promote a way of working which is sustainable' (Knightsbridge-Randall, 1996). Again there was a struggle to fund the project and no money available to share the experience with a wider audience.

The perspective of young black people was sought from a number of organisations: National Youth Agency; 1990 Trust; Society of West Indian Organisations; Sia; Black Environment Network (BEN). The Indian and African Caribbean Youth Project, part of the youth service in Oxford, writes 'as you will be aware black young people are rarely consulted about most things let alone about the environment' (e-mail response to author from Indian and African Caribbean Youth Project 1996). It was pointed out that even on occasions when people do genuinely try and consult they find it difficult either to access the community or to establish a relationship, and the consultation is seen to be tokenistic. In a study of the cultural life of young people Willis makes the point that 'black young people use their cultural backgrounds as frameworks for living and as symbolic resources for interpreting all aspects of their lives' (1996a: 8). The youth project stated 'it is not that black youth are not interested in environmental issues, they are very clear that the greatest environmental issues are often linked to ancestral homelands. They are also clear that in Britain it is [in] inner cities, where they predominantly live, that some of the environmental focus needs to be' (e-mail response to author from Indian and African Caribbean Youth Project 1996).

My observation is that the environmental stage is framed in a core/periphery motif. A minority (the core) have the lead roles, the longest speeches and the highest profile, and the majority (the periphery) play supporting roles, relegated to the audience, and excluded from the scriptwriting. The peripheral frame is characterised by women, black people and young people.

If the role of youth is to be advanced and young people are to be actively and creatively involved in a sustainable future then there are a number of constituents within the Agenda 21 frame that need to reassess their role. These include: NGOs, local authorities, government, academics and conservationists/environmentalists. The Development Education Association undertook a research project on development education and youth work in January 1995 which revealed the concern that young people may be seen as a potential group for raising money or helping the agency have a higher profile via campaigning. The specific needs of young people were often low on the NGOs' agenda (Bourne and McCollum, 1995).

Talking about future generations, Agenda 21, the move towards the end of the millennium and the importance of ensuring a sustainable world have become topics of conversation in many circles. I have pointed out elsewhere (Knightsbridge-Randall, 1996; 1998) that the language of environmentalists can act as a barrier to many individuals and groups. Terms such as biodiversity, capacity building, sustainability and indicators are not understood by the vast majority of young people, and while they may be interested in the environment they are excluded by lack of knowledge of a language that has no relevance to them. Further confusion is caused when politicians talk about the economy being sustainable. Nothing is more apt both to encapsulate and therefore isolate a group as linguistic barriers.

> Environmental organisations have essential lessons to learn in order to be able to understand the process of enablement and to begin to play a role in creating a real structure of support, training and funding for what they want to achieve. It means expanding beyond the environmental field in order to get the environmental elements going. Environmentalists must have faith that their investment will repay them . . . with vast numbers of aware and active people, who live lives which contribute to sustainability.
>
> (Ling Wong, J., Director of the Black Environment Network 1994)

To increase its impact and longevity, O'Riordan (1991) urges that the environmental issue should lose its distinctiveness and become embedded in other issues, such as the economy, education, defence and foreign policy. Academics might also consider the audiences that they do not reach and why this is so. Usher (1996) writes that research is necessarily embodied in the production of a written text. He quotes Parker and Shotter (1990) in his discussion on textuality and reflexivity who point out:

> academic texts by the use of certain strategies and devices, as well as predetermined meanings [are] able to construct a text which can be understood (by those who are party to such moves) in a way divorced from any reference to any local and immediate contexts. Textual communication can be relatively decontextualised.
>
> (Parker and Shotter 1990: 2)

Usher states that 'academic texts work in ways that make them appear as if they are located in no particular context' (1996: 43). He uses the word 'appear' because their context is actually the research practices of the relevant community. He argues that reading such texts requires initiation into the community and having command of its shared predetermined meanings. It requires, as Parker and Shotter emphasised, being party to the necessary moves. For those who are not, the text is meaningless – they literally do not have the means to enable them to read it.

The values and attitudes, skills and behaviour consistent with sustainable development are often practised by individuals and groups, but the connection to Agenda 21 is often not made. The intention to involve the community murmurs through each LA21 publication like a mantra of best intentions. LA21 usually comes under the domain of environmental services or the equivalent department. Poor communication between council departments results in local authorities not drawing in the educational and community development skills of other departments.

A survey of WWF's partnership work with local authorities on LA21 showed that the impact of sustainability principles on social policy issues mirrored the findings of the Local Government Management Board's survey (Tuxworth and Thomas, 1996). Both surveys revealed an accent on traditional environmental activities, land-use planning, waste management and energy management. Connections need to be made between professionals, but obstacles exist in many shapes and forms. The culture of local authorities and the background of professionals does not lead naturally to co-operative creative ventures. Another reason for the continued marginalisation of LA21 might be the fact that local authorities have a mechanistic demand of its services. Landry and Bianchini (1995) point to the fact that public officials are accountable to electorates and have set procedures to follow. There is also a concern that open policy development will produce an abundance of ideas and expectations that cannot be met. They observe that bureaucratic mind sets frequently imbue the whole organisation. Complex rules and regulations such as planning permission, licences, bye-laws and traffic restriction are needed to keep the machine running. Landry and Bianchini say that rules of this kind tend to be long-lasting, and for good reasons, resistant to change. Another view is 'if it ain't broke don't fix it'; which means that issues are only addressed if they become problems (Landry and Bianchini: 20).

While evaluating the implementation of a council policy in the 1980s (Knightsbridge-Randall, 1994), I cited comments from a book on educational change (Fullan, 1991). It is my view that the sentiment and advice offered are necessary for the journey to a sustainable future. Fullan said that 'the challenge of reform is not simply to achieve the implementation of single innovations' (199 : 29). He referred to Cuban's (1988) categorisation of innovations into *first-order* and *second-order* changes. First-order changes are those that improve the efficiency and effectiveness of what is currently done, 'without disturbing the basic organisational features' (1988: 342). 'Second-order changes seek to alter the fundamental ways in which organisations are put together, including new goals, structures, and roles (e.g. collaborative work cultures) . . . the challenge of the 1990s will be to deal with more second-order changes' (1992: 29). The challenge for the twenty-first century is to think radically in terms of second-order changes. The question remains: does a unity of vision exist to achieve this?

A new youth approach

Jeffs and Smith (1994) argue the need for a democratic audit of all youth policy and programmes, to ask in whose interest they are constructed?, and to apply an analysis that commences from a recognition of Berger and Luckman's (1967) principle that those who define reality are those with the biggest stick. They suggest that it would be a useful starting point if agencies and workers were to ask of all initiatives, do they:

- enable all to share in a common life?
- encourage people to think critically?

- foster the values and attitudes of a free society?
- sustain and extend opportunities for political participation?
- contribute towards greater equality?

(Jeffs and Smith, 1994: 29)

At an international conference on Youth at Risk, Hall (1995) commenting on initiatives presented by correspondents from Europe, asked the audience 'will the work be empowering young people at risk to regain some control over their lives in a future reality that they will inhabit, rather than the world we have known?'. Willis (1996b: 10) points out that young people's lives are full of expressions, signs and symbols through which they establish their presence, identity and meaning. The response to the question why focus on youth? is that:

> the teenage and early adult years are important from a cultural perspective in special need of a close 'qualitative' attention because it is here, at least in the first-world western cultures, where people are formed most self-consciously through their own symbolic and other activities. It is where they form symbolic moulds through which they understand themselves and their possibilities for the rest of their lives. It is also the stage where people begin to construct themselves through nuance and complexity, through difference as well as similarity.
>
> (Willis, 1996a: 7)

A new vision for youth

Involves:

1. A holistic view of young people and their environment that allows for the diversity of youth experience.
2. The encouragement of process and practice not product.
3. The building of partnerships where participants are facilitators and catalysts;
4. A distinction between passive participation and active participation.
5. The active involvement of young people in the development process.
6. Opportunities for young people to speak out in their own voice and be seen and heard.
7. The strengthening of community support networks.
8. The recognition and responsibility of all service agencies, institutions and organisations to involve young people at all levels of decision making.
9. A greater use of the human and physical resources of universities, colleges, schools, community and municipal buildings.
10. Funding that can be accessed by young people.
11. Working with the media and other information networks to raise the awareness of the real situation of young people and provide opportunities for young people to have access and opportunity to present their views, ideas and solutions using a range of styles and formats.

12 Encouraging fresh perspectives, creative minds and a capacity to explore other points of view.

ACE (Act, Create, Experience)

The Youth and Agenda 21 Report (Knightsbridge-Randall, 1996) recommended that WWF draw on its skills, expertise and reputation to support youth participation in Agenda 21 through an ACE programme that involved action, creativity and experience. The ACE approach was designed as a learning process involving awareness, experience and skill development. The opportunities suggested for young people were:

- International links
- Permaculture/urban food growing
- Multi-media/video projects/World Wide Web
- Events/conferences/training schemes
- Information
- Music projects
- ACE action

The WWF pilot project undertaken through 1996–7 revealed that the imagination of young people can be captured and released in the provision of lively, challenging, creative opportunities (Knightsbridge-Randall, 1998).The ACE agenda for young people's participation in the Earth Summit embodies the spirit and message of the new vision in this chapter and chimes with Willis's (1996a: 37) premise for programming policy which 'provides opportunities for young people to speak and represent themselves in their own voice and style' (1996a: 37).

Globally there is a recognition that environmental and social issues are intertwined. The words and expressions of Katy from Wales and Samantha from North Yorkshire speak for the youth agenda they have developed (Knightsbridge-Randall, 1998). (If you are curious you can join them in cyberspace at http://www.envirovision.org.)

> A further good point concerning the Agenda 21 Conference was that it made people think. There was none of the usual we're running out of coal so we can save energy by switching lights off, while the audience sat back and let all the information go in one ear and out of the other! We were made to think, to use the information we had gathered. We were made to offer solutions by telling ourselves, there's this, this and this wrong with the way we are living. What are we going to do about it and how are we going to do it? So for once, in contrast to the usual scenarios, there were few empty promises about doing more recycling and putting food out for the birds, and more positive statements and experiences that gave us the opportunity to make the connections and left a lasting impression.
>
> (Katy Broadhurst, South Wales)

> I began Powerhouse in conjunction with Project 21 because it was based on the environment. I knew nothing more than it was to run for 10 weeks and it involved the community, so, I really didn't know what to expect. Powerhouse began as another out of college hours activity and escalated to a thought provoking, emotional duty which woke me up to make me realise Yes! I can think how brutal it is to hunt whales, selfishly chop down the Rain Forest and not plant more and inhumanely experiment with nuclear weapons in the oceans. But what would I achieve from thinking? And so, to present Powerhouse to the public and most of all to the politicians, who had no room in their diary to attend, was an achievement and I am glad I was one of the team.
>
> (Samantha Kenny, North Yorkshire)

And so we come full circle, the message for adults *is* simple: don't believe your own prejudices. Youth will reach out and they want to be heard, do not let them down.

Notes

1 *Media Guardian*, March 3 1997, reporting on a MORI survey on the media, conducted for City University during 1996. Article written by Linda Christmas, senior lecturer in journalism at City University and Robert Worcester, chairman of Mori, and visiting professor at City.

2 A straw poll conducted at a Ditchley Foundation conference for top people from the British and foreign press showed that participants greatly under-estimated the number of young people regularly reading newspapers; the figure is 64 per cent. The research also showed that under-25s want to read about international as much as home news.

References

Barthes, R. (1964) (trans.) *Elements of Seminology*, London: Cape.

Barthes, R. (1975) *S/Z* (trans. Richard Miller) New York: Hill and Wang.

Benedetti, R. (1990) *The Actor at Work*, Englewood Cliffs NJ: Prentice-Hall.

Benetello, D. (1996) *Invisible Women: Detached Youth Work with Young Women*, Leicester: Youth Work Press.

Berger, P. and Luckman, T. (1967) *The Social Construction of Reality*, Harmondsworth, Penguin.

Bourne, D. and McCollum, A. (eds) (1995) *A World of Difference*, London: Development Education Association.

British Youth Council (1995) 'Politics and voting: information on issues affecting young people', briefing leaflet, London: British Youth Council.

Children's Rights Development Unit, in association with UNICEF, UNA and Calouste Gulbenkian Foundation (1993) *Background to the Convention*, London: CRDU.

Coles, B. (1995) *Youth and Social Policy*, London: UCL Press.

Cuban, L. (1988) 'A fundamental puzzle of school reform', Phi Delta Kappan 70(5): 34–44.

Eade, J. (1997) *Living in the Global City*, London: Routledge.

Elam, K. (1991) *The Semiotics of Theatre and Drama*, London and New York: Routledge.

Fainstein, S. and Campbell, S. (eds) (1996) *Urban Theory*, Cambridge MA and Oxford UK: Blackwell.

Fullan, M.G. (1991) *The New Meaning of Educational Change*, London: Cassell.

Gulli Pugliatti, P. (1976) *I Segini Latenti: Scittura Come Virtualita in King Lear*, Messina and Florence: D'Anna.

Hall, P. (1995) 'Youth at risk', keynote speech at the International Community Education Association (ICEA) conference, Edinburgh.

Harland, J., Kinder, K. and Hartley, K. (1995) *Arts in Their View, A Study of Youth Participation in the Arts*, Slough: National Foundation for Educational Research (NFER).

Harrop, J. (1992) *Acting*, London: Routledge.

Jeffs, J. and Smith, M. (1994) 'Young people, youth work and a new authoritarianism', *Youth and Policy* 46: 17–30.

Jones, G. and Wallace, C. (1992) *Youth, Family and Citizenship*, Milton Keynes: Open University Press.

Kempton, W., Boster, J. S. and Hartley, J. A. (1996) *Environmental Values in American Culture*, Cambridge MA: MIT Press.

Knightsbridge-Randall, S.J. (1994) 'The Drama of a Policy Process', unpublished PhD thesis, Cranfield University.

Knightsbridge-Randall, S.J. (1995) 'Youth at risk in England and Wales' in *Community Education Responses to the Needs of Disadvantaged and Excluded Young People across Europe*, ICEA: Association Nationale pour des Espaces D'Intergration.

Knightsbridge-Randall, S.J. (1996) 'Developing Agenda 21 initiatives with youth/young people', Internal Report, Godalming: WWF (UK).

Knightsbridge-Randall, S.J. (1998) *ACE, Youth and Agenda 21*, Godalming: WWF (UK)

Landry, C. and Bianchini, F. (1995) *The Creative City*, London: Demos.

Lawrence, G. (1996) *An Agenda 21 for London*, London: Association of London Government.

Ling Wong, J. (1994) 'Having faith in the community and ourselves', *NEST Magazine*, London: NCVO.

Lodge, D. (1977) *The Modes of Modern Writing, Metaphor, Metonymy, and the Typology of Modern Literature*, London, Melbourne and Auckland: Edward Arnold.

O'Riordan, T. (1991) 'Stability and transformation in Environmental Government', *Political Quarterly*, 62(2): 167–85.

Parker, I. and Shotter, J. (eds) (1990) *Deconstructing Social Psychology*, London: Routledge.

Parsons, W. (1995) *Public Policy*, Aldershot and Brookfield, Vermont: Edward Elgar.

Quarrie, J. (ed.) (1992) *Earth Summit '92: The United Nations Conference on Environment and Development*, London: Regency Press.

Sarason, S. (1990) *The Predicable Failure of Educational Reform*, San Francisco: Jossey-Bass.

Scott, D. and Usher, R. (eds) (1996) *Understanding Educational Research* London: Routledge.

Smith, D. E. (1987) *The Everyday World as Problematic*, Boston: Northeastern University Press.

Tuxworth, B. and Thomas, E. (1996) *Local Agenda 21 Survey*, London: LGMB.

United Nations Centre for Human Settlements (HABITAT) (1996) *An Urbanising World*, Oxford: Oxford University Press.

United Nations Environment Programme (UNEP) (1995) *Poverty and the Environment*, Nairobi: UNEP.

Usher, R. (1996) 'Textuality and reflexivity', in D. Scott, and R. Usher (eds) *Understanding Educational Research*, London: Routledge.

Willis, P. (1996a) *Common Culture*, Buckingham: Open University Press.

Willis, P. (1996b) *Moving Culture*, London: Calouste Culbenkian Foundation.

8 Including women: addressing gender

Susan Buckingham-Hatfield and Judith Matthews

Introduction

In 1992, the United Nations Commission on Environment and Development agreed Agenda 21, a programme for global sustainable development. This was a significant achievement since Agenda 21 was debated at the instigation of the NGOs. Signatories committed themselves to depositing a national plan for sustainable development by 1994 and local areas were required to produce local strategies by the end of 1996. One of the guiding principles of Agenda 21 is that people normally excluded from the decision-making process (such as women, indigenous people and young people) need to be integrally involved in decision making within a framework which stresses the importance of public participation. The reason for this inclusive form of participation is that these under-represented groups are seen as having had little impact on the production of environments, although they are sometimes disproportionately affected by them. Therein, however, lies a problem, as the structures which traditionally exclude these groups are being invoked to involve them fully. Moreover, greater participation needs a social structure which fosters and encourages such involvement, addressing concepts such as citizenship and empowerment, availability of information, education and a respect for people's identification with place – and place identity – which is in turn affected by environmental problems. The scene is then set for dissonance between a global agenda heavily influenced by NGO input and its national and local incorporation through political structures. This chapter examines aspects of the extent of this dissonance at national and local level in Australia and the UK, and seeks both to draw comparisons and to identify lessons which can be learnt from their experience.

The Australian and UK governments both adopted a strong public position on the development of Agenda 21 strategies following the UNCED Conference in Rio, and since both operate planning and public consultation procedures within similar legislative frameworks, a comparative perspective offers the opportunity to identify consequences which are structural, rather than consequent upon the particular administrative system. Environmental differences between Britain and Australia present communities with different priorities relating to physical environmental sustainability. For example, the environmental problems encountered by the UK are influenced by high population densities, and focus on issues such as urban congestion, population pressure on the countryside, waste disposal and transport. In Australia, on the other hand, a sparse population is concentrated in widely separated urban centres in a predominantly arid or semi-arid climate, and environmental issues accordingly focus on problems of soil degradation and water scarcity, and have quite different emphases as between the urban areas and the remote outback. The two countries nevertheless share the problems associated with advanced capitalism such as over-consumption and socio-environmental inequalities between rich and poor. Comparison between them allows us to explore the degree to which observed differences in the nature of concern, as well as the means of taking action, between men and women, are consistent across different environmental conditions, and to explore some of the ways in which their participation may be more effectively facilitated. In both countries, attention is directed toward the mechanisms by which public participation is being developed, with a particular focus on the experience of participation in the case of the UK. In Australia, the focus is directed more toward the approach adopted by female and male planning officers to the task of promoting such participation.

The UK and Australian government response to Agenda 21

Chapter 24 of the United Nations Commission on Environment and Development's Agenda 21 (United Nations, 1992) requires signatories to Agenda 21 to raise the capacity of women to participate in environmental decision making and to ensure that structures of decision making facilitate this. This entails a wide programme to promote literacy, health care, child care, equal opportunities, the ending of discrimination in both the public and the private spheres, and the portrayal of women in a positive way. The document recognises that women and children are particularly vulnerable to environmental damage and their perspectives need to be incorporated in environmental decision making by the active and equal participation of women at all levels. Given this unequivocal commitment, it is instructive to consider how the governments have responded in their submissions to the United Nations outlining how they plan to achieve sustainable development.

In the UK *Strategy for sustainable development* there is one mention of women; in the section on NGOs, the authors claim that 'areas in which NGOs work particularly well include primary health care, the needs of women . . .' (HMSO, 1994: 195). Even in sections which discuss population, household formation and income, any gendered perspective is missing, as it is in *Putting Sustainability into Practice* and in the government's own *Principles.* To illustrate the distance the UK government has to travel to come within sight of the United Nations' principles: on fertility 'The Government believes that couples should make their own decisions about how many children to have . . .' (HMSO, 1994: 8); this presupposes that all parents are part of a 'couple' and marginalises a woman's right to determine her own fertility. In his introduction to the document, the then Secretary of State for the Environment opens his statement with 'Man has grown used to living as conqueror . . .' (HMSO, 1994). There is an observable trend that government (local as well as central) is transforming citizens into consumers, but even in the more protracted discussions on consumers, there is no acknowledgement that consumption activity may be gendered.

In the 1996 *Common Inheritance* document, participation is not mentioned at all for women. In the international context, there is reference to the Overseas Development Agency supporting programmes to improve the health and education of women in the South, to improve the economic, social, political and legal status of women, and offering child allowances to enable women from the South to attend training in the UK (HMSO, 1994: 371–3). There is, as a consequence, nothing in the commitment tables about women (or participation) in this UK document.

The Local Government Management Board, who support local authorities in the development of local environmental agendas, published guidelines on including women, emphasising consultation and participation rather than decision making. The document (one of the last guidance notes to be produced, along with the pamphlets on involving young people and ethnic minorities) rather patronisingly assumes that women need to be taught their environmental responsibilities, when we would argue that women are those most likely to be aware of the environmental implications of their actions and that they are frequently restricted in their ability to respond because of structural relations. Compared to the examples from South Australia later in this chapter, the model case studies are the necessary, but rather limited and far from innovative examples of nappy washing schemes, environmental consumer advice and organic food co-operatives (LGMB, 1996).

Australia's response to Rio was the development of a *National Strategy for Ecologically sustainable development*. This is known as the ESD strategy. The strategy document (Commonwealth of Australia, 1992) outlines the 'challenges', 'strategic approach' and 'objectives' to be pursued in relation to sustainable development for each of the sectoral and inter-sectoral issues which it identifies as elements in the pursuit of sustainable development. Its section on 'gender issues' identifies the challenge as being that of developing 'ESD related policies, programs and actions which incorporate the particular concerns of women, while ensuring that actions to achieve ESD do not have inequitable effects on women'. Strategically, state governments were to take action to ensure women's access to information and decision-making processes, and this was envisaged as involving the need for governments to recognise women's interests in the development of nationally co-ordinated public education and information to improve awareness of ESD issues, providing information and seeking input from women's representative groups, ensuring women's participation in co-operative approaches to resource management issues, and ensuring participation of women on ESD-related decision-making bodies and advisory groups. In the 1995 Annual Report of the Commonwealth Government to the United Nations Commission on sustainable development, the section relating to women notes that 'Australian women see environmental issues as a major concern'. However, the report notes that this importance, and the high profile that women have in environmental community groups and in the debate on environmental issues 'is not yet adequately reflected in many of the decision making processes which have significant impact on the environment' (Commonwealth of Australia, 1995: 54). The report goes on to outline various undertakings to improve participation by women, and notes the changing emphasis on 'gender' issues in work by development agencies. However, a review of the listing of Ministerial Media Releases from the Environment Minister's office since the election of the new national government in February 1996 shows that of the 219 such statements recorded at the time of writing (July 1997), none mentions women in its title (Australian government on-line Environment Information Service at www site-http://www.environment.gov.au/portfolio/minister/env/96.html, and /env/97.html).

The role of citizenship

Traditionally, the concept of citizenship has been bound up with the creation of nation states which involved the shift from subject status to emancipation based on ideas of universal rights, freedom of expression and political liberty. Geoff Andrews (1991) suggests that there is currently a rebirth of interest in citizenship which is taking place against the decline in importance of the nation state, and that this will alter the international political landscape. The value of the nation state is also brought into question with regard to its capacity to deal with environmental problems which are inherently transnational (see, e.g., Princen and Finger's (1994), on NGOs). Indeed, the recasting of the European Community by the Maastricht Treaty (1992) entitles all nationals of Member States to claim European citizenship which guarantees freedom to move within the Union and to enjoy economic and political rights anywhere in the Union. However, citizens' rights and responsibilities are increasingly tied to economic status and, as we have already indicated this facilitates the transformation of citizens into consumers. This is particularly true in the case of women who are repeatedly framed in terms of consumption. Garcia Ramon and Monk (1996) show clearly how citizenship in Europe is linked to economic independence and the effects this has on women's access to resources. Although Lipietz (1995) is scathing about Maastricht's capacity to facilitate environmental co-operation, he argues persuasively that a Europe without political frontiers is important if this is to be achieved.

The effective exercise of these rights is constrained by a citizen's means to claim and enjoy the rights and to accept responsibilities. Andrews cites Ruth Lister who believes that 'active citizens' are those 'able to stand alone, independent before the market, their freedom guaranteed by economic rather than social rights', suggesting a prioritisation of the consumer (Lister in Andrews, 1991: 13). The exercise of rights and responsibilities also depends on information. In some cases a lack of information may prevent people from playing a full part in civic life, in others it may be that the information and knowledge that people have to offer is not recognised by decision makers. Elsewhere, Lister (1996) argues that citizenship itself has a gender dimension, suggesting that, by its emphasis on participation in public life, women are often excluded from full citizenship. Reinforcing this is the ambiguity of rights in the private sphere and the 'right' to citizenship exercised through participation in paid employment. The same argument can be extended to children and to others not in paid employment and confined to the private sphere, such as many of the elderly and people with disabilities.

Women are more active in what Vepsa and Horelli (1995), following feminist Scandinavian practice, call 'intermediary space' – between the public and private sphere. It is well documented (Garcia Ramon and Monk, 1996; HMSO, 1994a) that women take the overwhelming burden of domestic work, even though they may be involved in paid work in the public sphere. Waring (1988) eloquently argues that the devaluation of women's domestic role renders them invisible, as it does in employment which is unregulated and/or least likely to benefit from representation by professional associations and trades unions.

Irwin (1995) argues that for the concept of citizenship to work effectively in practice, 'local' and 'citizen's' knowledge must be respected. If it is neglected, it will 'restrict the social learning between science, technology and public groups which . . . is essential to the process of sustainable development' (Irwin, 1995: 7). He goes further to suggest that if citizens are unable to take control of their own lives, health and the environment, then there will be no sustainability.

Local knowledges, of course, emerge from somewhere. Each person develops her or his views through experience which includes education – formal and informal – and exposure to the media. Local knowledge will also be affected by people's ability to acquire meaningful information and whether this knowledge is acknowledged may well depend on how it is expressed.

Place identity

Environment is experienced differently by men and women as a consequence of the different daily 'worlds' in which they operate. This difference in experience constitutes a significant element of what has been referred to as *place identity* – that element of a person's sense of self which derives from their experience of, and attitudes toward, the physical environment. An example of the implications of such differences is to be found in the contribution to the submission from the Office of the Status of Women to the preparation of the ESD initiative in Australia prepared by Brown (1992). She argued that women's uses of the environment are sufficiently different from those of men as to constitute a distinctive 'habitat', in the ecological sense. Such a position echoes that proposed by Mellor (1996) in arguing that women are more embedded in the environment, as a result of their greater involvement with the work of nurturing and caring. This results in an awareness of the impacts of environmental damage which needs to be acknowledged and developed by both women and men in the process of working toward sustainable development. In the UK, research in West London has demonstrated that women's concern for environmental issues is consistently higher than is men's (with the exception of concern about run-down buildings). Women with children are more likely than those without to be concerned about local environmental problems and place identity is likely to play a role in this (Buckingham-Hatfield, 1994).

The differences in power between men and women lead not only to differences in experience and, in many cases, to greater risk and greater awareness of the immediate and local risks, but also to differences in the capacity to name the risks which should be the priorities. Brown and Ferguson (1995) present a particularly compelling case for paying much greater attention to the particular roles of women in activism focused on environmental risk. In a survey of organisations and members of groups involved in actions over toxic waste in the United States they point out that women constitute the majority of both the leadership and the membership of local toxic waste activist groups. In spite of this, they argue, gender and the fight against toxic hazards are rarely analysed together – either in studies on gender or on environmental issues. Their analysis concludes that toxic waste activism constitutes a burgeoning social movement which is organised around local problems (particularly where these threaten health), is concerned with an approach to knowledge which is democratic and collective, and synthesises subjective and objective features. The now classic case of Lois Gibb's approach (exhaustively referred to – and see also Gibbs 1998) to developing a linkage between her own subjective awareness of the health problems at Love Canal, and the incorporation of a scientist's ability to present data in the form necessary to persuade the authorities, is shown in this analysis to be reflected again and again in the groups which Brown and Ferguson studied.

Identification with place mobilises the nature of response to risks. It does so through the way in which it taps or impacts upon people's 'place identity'. Women express their identification differently, but if their expressions are not given voice in LA21 or other planning and management processes, then changes which result from such actions are highly likely to leave women's concerns unincorporated, and their risk exposure unchanged. More importantly, the impact of such changes is highly likely to generate new place identities for participants which reinforce exclusionary tendencies in environmental negotiation by failing to incorporate perspectives that derive from women's concerns. If new place identities are forged in ways that continue to exclude the perspectives of women and other under-represented groups, then conflicts over resources and resource use and over the relationships between social and environmental issues will maintain many of their current characteristics. Furthermore, by maintaining existing inter-group distinctions based on gender as a normal element of prevailing representations of the 'environmental agenda', the potential for developing support for practical alternative approaches which are present in the viewpoints adopted by women will be lost. This will be a different kind of loss, in that it will derive both from failure to take note of their concerns, but also from the loss *to women* of access to such perspectives. The woman whose place identity is forged in the context of an environmental agenda which discounts women's perspectives will become less likely to recognise the validity of such perspectives – until another Love Canal situation emerges – and perhaps not even then.

Public participation and LA21: the local dimension in West London

A study of the participation process for LA21 in West London highlights the degree to which identifications with the locality resulted in different perspectives on its needs as between women and men, and also demonstrates the extent to which the organisation, location and running of participation processes generate differences in access to citizenship rights between them.

Drafting the consultative document

The primary research was undertaken in the London Borough of Hounslow, through participant observation, documentary analysis, questionnaires and interviews. Documentary records of the process in neighbouring boroughs have also been considered. Hounslow encompasses an area affected by major trunk roads and motorways and by Heathrow Airport, and the borough is, at the time of writing, involved in opposing the fifth airport terminal proposed by British Airports Authority. The borough's environmental opportunities include a stretch of the River Thames and a number of tributaries and several major open spaces.

In Hounslow, the LA21 process was inaugurated in May 1995 at an open meeting on a Wednesday evening (7–9p.m.) at the Borough Civic Centre – a large, sprawling centre, heavily landscaped with trees and bushes through which footpaths meander attractively, if also potentially threateningly. Around 60 people attended a talk from the councillor responsible for environmental issues and the borough's adviser (a paid consultant). After this meeting, the participants aligned themselves to a group closest to their interests: three were formed that evening (Transport and Air Pollution, Waste Disposal and Recycling and Greenspaces), a further two formed a few months later (Energy Use and Water). Each group was 'facilitated' by a council officer and another officer took notes of the meeting and compiled a contacts list.

The Transport and Air Pollution Group, with which Sue Buckingham-Hatfield worked, attracted 20 participants (13 men and seven women), was served by three (male) council officers; a (male) chair and vice-chair volunteered and were 'elected' unopposed. At this stage, two factions could be identified within the group, although not everyone fitted these stereotypes: younger men who tended to represent cycling groups and older women, who were more concerned with issues of environmental incivility such as noise, exhaust fumes, litter and graffiti. Whilst these two sets of concerns are not mutually exclusive, the terms of the dialogue were set early on, with guidance from the Chair, between the importance of *sustainability* and the daily concerns of noxious behaviour – with the latter concerns sometimes being trivialised and judged inappropriate for LA21. Sustainable development as a concept is still not widely recognised (and is highly contested amongst those who do recognise it, see Buckingham-Hatfield and Evans, 1996), and this can have an alienating effect upon those who are not familiar with the term. Its use can legitimate and prioritise contributions in the environmental debate and minimise the importance of local or parochial terms because they are not framed in the 'appropriate' language of futurity. It also raises the question as to which, of conflicting public views, gets incorporated into formal plans.

The group met a further five times prior to the drafting of the consultation document; three of these meetings were at the Civic Centre, the last two, because of the unavailability of rooms there, at a local council run centre at street level off the high street. Nonetheless, attendance at the meetings steadily dwindled with those participants not returning least likely to fit the stereotypes mentioned above. At one meeting before Christmas, the number of officers outnumbered the members of the public.

A record of the process indicates that interest in the meetings waned since the third, when the officers took quite a strong lead in directing the meeting to form workshops which were expected to produce the draft of the consultative document. The deference shown to the experts or authority figures in some ways interfered with the dynamics and initiative of the group and the remaining active members were either regular participants in local authority fora (known to the officers), officers themselves or environmental professionals. It was the latter group (two university academics and an employee in the environmental profession) who undertook to draw up the consultative document which was presented to an open meeting in May 1996, again in the Civic Centre.

The process of consultation

The draft document was presented to an open meeting, better attended than any of the previous meetings. The contributions were generally well informed and a number of issues were raised which participants felt had not been adequately covered in the document. It was felt that Heathrow Airport needed a higher profile; some derelict and defunct marshalling yards designated environmentally sensitive had been scheduled for redevelopment, which was causing some concern, and there was a feeling that social equity issues were not given enough prominence.

Following this meeting, 3,500 copies of the draft documents were distributed via schools and libraries and directly to around 300 individuals and organisations, although only 100 replies were received. It is extremely difficult to draw any conclusions from such a small number of responses, but a number of points can be made which raises the question of equity of participation. It was noticeable that almost half of the respondents represented community and voluntary groups (41 per cent), a quarter were from individuals and the remaining 36 per cent were from businesses, statutory agencies, transport interests, a school and a political party. Thirty five per cent of responses came from outside the borough, which raises interesting questions about representation.

Not unexpectedly, almost twice as many men responded as women (58 per cent compared to 30 per cent of respondents, the rest were unidentified). Women were most likely to be individuals or representing environmental or heritage groups whilst men were more likely to represent institutions such as statutory agencies and companies, both of whom were exclusively represented by men. This separation of interests reinforces the divide based on expertise between women and men.

On a whole range of concerns cited more than once, women expressed greater concern: 78 per cent of women responding thought that transport was an issue compared to 49 per cent of men; 38 per cent of women agreed that waste disposal was an issue compared to 16 per cent of men; on green space, 30 per cent of women respondents compared with 18 per cent of men expressed concern. On air pollution, 13 per cent of women compared with 4 per cent of men named it as a problem, and on energy consumption the levels of concern were 13 per cent for women and 7 per cent for men.

Women tended to be more constructive in their suggestions. For example, they were more likely to suggest additions to the draft document. Of the suggestions made there appeared to be a gendered emphasis with women referring to environmental education, community participation, tree-planting and recycling, whilst men, in addition to citing community participation, focused on auditing, publicity and cycling. Interestingly, no women cited cycling as an issue which, in the section on transport concerns was exclusively a male interest; conversely, only women referred to concern for pedestrian priority here.

Women were more likely to comment on the interpretation of the document (17 per cent compared to 7 per cent of male respondents), on the layout (78 per cent of women compared to 47 per cent of men) and showed more concern over the likelihood of implementation once the document has been ratified (22 per cent of women compared to 11 per cent of men). Furthermore, women were proportionately more interested in becoming involved (30 per cent of women, compared with 24 per cent of men were interested in joining a working party) and 65 per cent of women compared to 53 per cent of men asked to be kept informed of developments.

This, however, is at odds with the actual participation rate of women in workshops; follow-up interviews were conducted to explore this apparent contradiction. These suggested that 'lack of time' is the most significant barrier to more active involvement in Local Agenda 21. Since women generally have less leisure time than men (Droogleever-Fortuijn, 1996), this is likely to be one reason for the lesser involvement of women. Venue and time were cited as problems, but less often.

Local conclusions

Two factors have prevented fully participatory discussions taking place on transport and air pollution (the working group which was monitored). One is the organisational structure in which meetings take place in the evening in council offices, which Booth (1982) long ago claimed was destined to exclude large numbers of women from public participation. Interestingly, she continues to cite this as relevant in her latest (1996) publication, which suggests there has been little change in the last 14 years. Such locations have other disadvantages such as the lack of accessibility for residents in other parts of the borough – why not vary the location? The way in which the venue is organised is also off-putting; for example, for a two-hour meeting which may entail a total of one hour's additional travelling time there are no refreshments laid on. Why not provide drinks and light snacks, particularly for those coming straight from work? The second is the staffing and the control of the dialogue by mostly male, mostly 'expert', 'gatekeepers'. This aspect was most graphically illustrated in a conference held by a neighbouring borough to 'launch' the public participation aspect of LA21. Held in a centrally located adult education college on a Saturday, these organisational strengths were diminished by a list of eight speakers which contained only one woman (a local authority employee) and, in 13 workshops, no women facilitating and only one woman recording.

Planners, LA21 and processes of participation in Australia

Some useful contrasts with the West London experience emerge from observations made during the course of research designed to examine the adoption of approaches commensurate with LA21 in Australia. In late 1995, Judith Matthews visited a range of local government planning offices in the capital cities of four states (in South Australia, West Australia, Queensland and the Australian Capital Territory). Interviews were conducted with the planning officer who held responsibility for the environmental dimensions of the council's planning work. Of 12 officers, three were women, all of whom worked in South Australia. In addition, these were the only offices in which an approach was taken which explicitly used 'Local Agenda 21' as a label for its activities. Consequently, the research led to a fortuitous correspondence between an interest in LA21 processes and the desire to explore women's contribution to this process. No claim is made that this account is representative of Australia as a whole. The sample is very small, and the interviews were not designed explicitly to explore issues of difference between women and men in planning practice. Nevertheless, it is suggested that some illuminating perspectives arise from consideration of these particular cases, perhaps particularly because they emerged spontaneously from an enquiry couched more generally.

Australia's LA21 process has been managed from Municipal Corporations Association (now Environ) in Melbourne, and those local authorities where LA21 was being pursued used their guidelines quite strongly. The approach of MCA, outlined in *Managing for the Future: a Local Government Guide* (Cotter *et al.*, 1994) is presented in an attractively designed, glossy format handbook. This outlines the rationale for LA21 in accessible language, while drawing upon the detailed technical material to which local authority planners – the target audience for the publication – will have access. Having outlined the issues and the challenge to authorities, it is then very specific in outlining the steps which will need to be taken in developing a LA21 within an authority. Seven steps are identified, beginning with the establishment of a team and the recruitment of council support to proceed and culminating in the development of review and monitoring processes, sharing experience with other councils, and celebrating the outcomes of the process as a mechanism for engendering continued support and development since:

> it will never be clear or even possible to reach a final end point in moving toward sustainability. Understanding and sharing what each local area has achieved in terms of moving closer to sustainability goals should be celebrated because it will be a great achievement if the consistent themes of Agenda 21 are embraced.
>
> (Cotter *et al.*, 1994: 43)

The document is explicit at various points about the need for councils to take particular steps to include the concerns of women, young people and minority groups, and although it does not elaborate upon the mechanisms by which this is to be achieved, it is very clear that such interests must be valued. Suggestions for approaches to these groups include surveys in different languages, small group meetings in locations familiar to the group, forums, debates, festivals and the use of local media. Strikingly, the document argues that such approaches will result in commitment because of their power to 'engender a sense of identity within the community' and to foster a sense of place and of a locality's identity.

The women planners interviewed during the research who were developing environmental strategies using a LA21 framework shared a number of approaches. In each case, they emphasised the need to achieve a good spread of information and support within the council before going out to the community. Each emphasised the importance of participation and the equality of status of contributions, both in internal discussions and in participation by communities. There was general use of visioning methods, flip charts, and approaches that were not only based on words but on drawings, model-making and other approaches to the identification of local concerns. Information leaflets and response sheets were an important element of strategies to evoke participation by the community. Finally, there was a striking consistency in the use of multiple community venues for consultation exercises – not just the Council Offices, but also schools, community halls and private homes.

By contrast the male planners interviewed were not using LA21 as a framework. Many argued that their approach was framed according to the demands of the National ESD programme. This resulted in the development of strategies for particular environmental media, for example bush conservation, water management, catchment plans. Consultation involved setting up representative groups from local areas comprising specialists from the local community such as locally based botanists, chemists, or environmental technicians, and consulting the specialist environmental groups in their areas. An important part of their strategy was to prepare 'priority programmes', and issue these for local response, and to attempt to 'write in' environmental considerations to local plan systems. The male planners consulted during this research demonstrated a clear tradition of the kind of environmental management strategies to which the MCA Report gives credit, as highlighting the scientific dimensions of environment, and the big picture issues – greenhouse effects, conservation, biodiversity. What the Report advocates, and the women planners following LA21 approaches exemplified, was the way in which these issues need to be pursued in the context of a locally orientated focus which concentrates on achieving inclusive participation and commitment from the community before strategies were developed.

The Australian research happened to find women using the LA21 framework as an element of environmental planning strategy. This appears to confirm the tendency of women to approach the development of environmental programmes in a different way from that adopted by men. Further confirmation of such difference in the Australian context is to be found in the work of Costello and Dunn (1994) in a study which focuses not on the 'planners' but on the communities they serve. Their work examines the spatial spread and growth in numbers of Resident Action Groups (RAGs) in Sydney in order to examine the degree to which these predominantly represent NIMBY, regressive, middle-class groupings to protect favoured areas. They find that recently formed groups were more suburban and less likely to be located in affluent areas than the groups established at an earlier date. The groups' action focus is increasingly centred on environmental and urban development issues, by contrast with a stronger concern over the development of formal associations, and urban development themes in the earlier period. The action groups discussed in detail, which have been formed recently, are environmentally orientated, being concerned with the effects of chemicals pollution and waste disposal sites. These groups show a strong tendency to be led by women, they are effective in networking amongst themselves, and have given rise to attempts by local government to pre-empt protest by setting up consultative groups, which frequently are not drawn from amongst women. Costello and Dunn argue that the RAGs have the potential, through their forms of action and networking, to constitute a social movement. This leads to a situation in which the local industry and governments are increasingly forced to pay heed to the concerns of residents. The New South Wales government responded to this situation by seeking to curb the planning powers of local government, limiting the empowerment of RAGs – and because of the high involvement of women in the groups, therefore also of women.

> RAGs do not always make a serious challenge to the dominant order and are often single-issue groups with parochial aims. However, the view that RAGs are NIMBY rat-bag groups is simplistic. RAGs can provide a voice to those who would otherwise be disempowered. They can influence the decisions of the state and of capital. Whether they necessarily constitute an urban social movement depends on the definition of 'social movement'. If a social movement is an empowering force, in which women and men can be introduced to the political structures and become engaged in protest and resistance that can challenge the social and spatial order, then RAGs are social movements. After all, RAGs can, either singularly or cumulatively, force a realigning of existing power relations and, ultimately, necessitate changed modes of government.
>
> (Costello and Dunn, 1994: 74)

How then, are we to understand the social processes from which these differences emerge? It is important to note the role of identity, as a derivation from both individual experiences and group memberships and attachments in determining the processes by which women seek incorporation of their perspectives in decision making. A questionnaire survey by Kelly and Breinlinger (1995) directly explored this dimension of participation. The survey compared two groups of women. One comprised members of groups committed to action on women's issues, the other was drawn from women taking part in university postgraduate professional development courses who were not involved in women's issues groups. The respondents were interviewed twice, with a year between interviews. The study examined the relationships between participation in women's groups, willingness to engage in collective protest, informal participation and individual protest, and sense of relative deprivation, gender identity, efficacy, collectivist orientation in relation to gender ('women must act together') and general collectivist orientation ('I work better in a group than on my own'). The results show that the women who belonged to women's issue groups were significantly more committed to participation, protest, informal and individual protest. Participation in women's groups and in individual protest was related to higher levels of self-efficacy – the belief that one's actions can make a difference – so those who believed that getting involved makes a difference were more active than those who did not believe so. Those who were not involved in collective action were more likely to say that there is no point in doing so. No correlation was found between a sense of relative deprivation and participation. Those who reported higher levels of relative deprivation were more likely to engage in collective protest and individual action. The writers argue that this may be due to fact that protest and informal activities deal with general appeals to all women rather than the more issue-oriented focus of women's groups which is not motivated by generalised relative deprivation.

It could be argued that these findings are self-evident – that only those who believe action can make a difference will bother to become involved – but perhaps the key finding of this study was that although belonging to women's groups led to higher levels of participation, this was *not* strongly related to a belief in general collectivism ('I work better in a group than on my own').The writers of the study quote Klandermans (1984) in arguing that the significant dimension may be the social construction of injustice and protest. Disaffection with current political circumstances does not alone make people more active. Dissatisfaction has to be translated if it is to lead to social or political action. Group identification through gender may lead to different ways of responding to grievances for those strongly identified than for those with weaker group affiliation. Identification as an activist was the most important contributor to likelihood of actual involvement in groups or actions. To get people to become involved in actions it may be necessary first to get them to think of themselves as activists:

> To conclude, the present findings have pointed to the important role of identity processes underlying individual participation in collective action. Identification as a group member and as an activist should be incorporated

> into models of participation alongside those perceptions which have traditionally been the focus of attention, that social change is both desirable and possible.
>
> (Kelly and Breinlinger, 1995: 55)

Identity comprises individual components, for example the beliefs one has about one's personal efficacy in bringing about change, components of identification with, and separation from, particular groups, such as those which deliberately pursue women's issues, and it comprises attitudes toward the environment generally, and to particular places – the place component of identity. The ways in which women's identity is realised at all levels, including that of place identity, indicate the vital necessity of understanding these processes, and taking account of them in developing environmental programmes such as LA21.

Conclusions

The extent of gender bias in structuring the terms of the local environmental debate, let alone the processes for participating in it, is likely to have a significant impact on the extent to which women will engage with sustainability issues and shape future policy. Many commentators argue that the real impetus for sustainable development policy in the UK has come from local authorities (Agyeman and Evans, 1995; Buckingham-Hatfield and Evans, 1996), despite a lack of support from central government. This enthusiasm, however, has in many cases preceded any in-depth and incisive consideration as to how proper participation may take place. Few authorities, as this chapter illustrates, have employed techniques markedly different from those which have provoked little participation before and consequently it is difficult to see how the UN invocation to involve previously disadvantaged groups centrally in environmental decision making is to be materialised, and how, therefore, the environment is to become shaped by their needs. Nor is the good practice of the LGMB referred to earlier likely to inspire a change of practice. The Australian examples show how women planners have been able to use the LA21 framework as a means of developing strategies more inclusively. Both the British and Australian examples of public participation discussed here show how important it is to take account of identity processes, and of the differing aspects of local places which women and men highlight in their response to local environmental issues when developing policy documents and discussion papers. If such account is not taken, previously excluded groups, including women, will continue to see their concerns excluded from the decision-making process, as a direct consequence of the social processes that create place identities.

In some respects, despite a long tradition of civic activism, a well-developed feminist movement and environmental movement, environmental decision making in the UK is heavily dominated by experts, who are predominantly men. National policy declarations, such as *sustainable development: the UK Strategy* (1994) choose not to reflect on the gender dimension and, despite a local government tradition which acknowledges the importance of the women's voice, the

LA21 process appears to be soliciting public involvement in ways long held to disadvantage women and other groups and individuals ascribed low status. In Australia, by contrast, the national strategy contains a much stronger stated commitment to the need to incorporate women's concerns, although the evidence for particular recent initiatives is by no means overwhelming.

Other strategies which need to run parallel with increased local involvement such as effective environmental education and citizenship building are frequently ambiguously handled and yet are central to the LA21 framework, for unless sustainable development strategies are negotiated by the people who must implement them, they cannot succeed. LA21 requires the participation of citizens, and implies an extension to the empowerment of citizens. Whilst lip-service is paid to the concept by governments it is frequently not developed in ways likely to redistribute rights and responsibilities. Without it, a sustainable environment will not be achievable.

Women realise and demonstrate their environmental concerns in different ways than do men and also prioritise them in different ways. The social contexts within which these differences are demonstrated and within which they are formally interpreted, tend to diminish the importance of the alternative perspectives and approaches adopted by women. Attention needs to be paid to the means by which the alternative that women present can be incorporated into the mainstream of LA21. In the present discussion, the actions of the Australian women planners and activists are instructive. The issues themselves are not different, and the intended outcomes of the alternative approaches are common to men and women. Awareness of the processes of identification with place, and of the tendency of social grouping mechanisms to separate groups even where they share common interests, should focus attention on place as the basis of action, rather than social groups. By so doing, recognition of a common interest in local places can allow recognition that such common interest is undermined if particular groups are disadvantaged within it – hence women's, young people's, disabled groups' and many other interests can be addressed within a more collective framework of localised concern.

References

Agyeman, J. and Evans, B. (eds) (1994) *Local Environmental Policies and Strategies*, London: Longman.

Andrews, G. (1991) *Citizenship*, London: Lawrence Wishart.

Booth, C. (1982) 'Public participation and women' in Polytechnic of Central London (ed.) in *Women and the Planned Environment* conference proceedings, Polytechnic of Central London: London.

Booth, C. (1996) 'Women and consultation' in Booth, C., Darke, J. and Yeandle, S. (eds) *Changing Places: Women's Lives in the City*, London: Paul Chapman.

Brown, V. (1992) *Engendering the Debate*, report prepared for the Ecologically sustainable development Working Groups by the Office of the Status of Women and Department of the Prime Minister and Cabinet.

Brown, V. (1995) *Women and Environment Network (Australia)*, report.

Brown, P. and Ferguson, F.I.T. (1995) '"Making a big stink" : women's work, women's relationships and toxic waste activism', *Gender and Society* 9(2): 145–172.

Buckingham-Hatfield, S. (1994) 'Addressing environmental issues in the 1990s: a gendered perspective', *West London Papers* 2(1).

Buckingham-Hatfield, S. and Evans, B. (eds) (1996) *Environmental Planning and Sustainability*, Chichester: John Wiley.

Commission of the European Communities (1992) *Towards Sustainability, the Fifth Environmental Action Programme*, Brussels: CEC.

Commonwealth of Australia (1992) *National Strategy for Ecologically sustainable development*, Canberra: Australian Government Publishing Service.

Commonwealth of Australia (1995) *Australia's National Report to the United Nations Commission on sustainable development on the Implementation of Agenda 21, 1995*, Canberra: Australian Government Publishing Service.

Costello, L.N. and Dunn, K.M. (1994) 'Resident action groups in Sydney: people power or rat-bags?', *Australian Geographer* 25(1): 61–76.

Cotter, B., Westcott, W. and Williams, S. (eds) (1994) *Managing the Future a Local Government Guide*, Canberra: Department of Environment, Sport and Territories.

Droogleever Fortuijn, J. (1996) 'City and suburb: contexts for Dutch women's work and daily lives' in Garcia-Ramon, D. and Monk, J. (eds) *Women of the European Union, the Politics of Work and Daily Life*, London: Routledge.

Garcia Ramon, D. and Monk, J. (1996) *Women in the European Union*, London: Routledge.

Gibbs, L.M. (1998) *Love Canal. The Story Continues*, Gabriola Island: New Society Publishers.

HMSO (1994a) *Social Trends*, London: HMSO.

HMSO (1994b) *sustainable development: the UK Strategy*, London: HMSO.

HMSO (1996) *This Common Inheritance: UK Annual Report*, London: HMSO.

Irwin, A. (1995) *Citizen Science*, London: Routledge.

Kelly, C. and Breinlinger, S. (1995) 'Identity and injustice: exploring women's participation in collective action', *Journal of Community and Applied Social Psychology* 5(1): 41–57.

Klandermans, B. (1984) 'Mobilization and participation: social-psychological expansions of resource mobilization theory', *American Sociological Review* 489: 583–600.

Kofman, E. (1995) 'Citizenship for some but not for others: spaces of citizenship in contemporary Europe', *Political Geography* 14(2).

Lipietz, A. (1995) *Green Hopes: the Future of Political Ecology* Oxford: Polity Press.

Lister, R. (1996) 'Citizenship engendered' in Taylor, D. (ed.) *Critical Social Policy* London: Sage.

Local Government Management Board, (1996) *Women and sustainable development*, London: LGMB.

Mellor, M. (1996) 'Sustainability: a feminist approach' in Buckingham-Hatfield, S. and Evans, B. (Eds) *Environmental Planning and Sustainability*, Chichester: John Wiley.

Princen, T. and Finger, M. (1994) *Environmental NGOs in World Politics*, London: Routledge.

United Nations Commission on Environment and Development (1992) *Agenda 21*, Geneva: UNCED.

Vepsa, K. and Horelli, L. (1995) 'Women's networks as a strategy to integrate physical and social aspects in planning' in Ottes, L. *et al. Gender and the Built Environment* Assen, The Netherlands: Van Gorcum.

Waring, M. (1988) *Counting for Nothing*, Wellington: Allen & Unwin.

9 The North–South dimension of Local Agenda 21

Marek Lubelski and Raff Carmen

Introduction

> A sustainable community lives in harmony with its local environment and does not cause damage to distant environments – now or in the future. Quality of life and the interest of future generations are valued above immediate material consumption and economic growth.
>
> (LGMB UK 1994a: 12)

> What would happen if the whole population of the South devoured the world just as unpunished as the North and with the same voracity?
>
> (Galeano 1992)

This chapter affirms that the challenge of finding creative and practical responses to the process and direction of global development, now dubbed globalisation, needs to be a basic element of our working towards more sustainable development in the coming years. At the same time, it emphasises the dynamics of the credibility gap in view of Agenda 21's (A21's) incongruous status within the context of the dominant Northern vision of the future currently driving the overall agenda of international policy-making and its outcomes. It then goes on to identify some characteristics of a range of responses and initiatives predating and coinciding with Agenda 21, which emphasise alternative, Southern perspectives, and address this context of globalisation by contributing to an alternative, citizens' agenda for communities and their environments in the twenty-first century.

It is a manifest contradiction that the strategic Agenda 21 document of the UN summit in Rio which comprises no less than 40 sections, should be a *statement of intent*, the ultimate 'Blueprint to Save the Planet', by the signatory governments professing commitment to a cleaner, fairer and safer world. The text belies the real agenda to which the signatories are, *de facto* and ever-increasingly, dedicated; namely, an overarching allegiance to *unsustainable international development priorities* premised on the imperatives of economic growth, market liberalisation, and the propagation of the pseudo-political consumerist culture of freedom of choice. From within this edifice are heard the relentless mantras of continually discredited rhetorical solutions to the ills of societies the world over; namely, deregulation, privatisation, efficiency, flexible human resources and so on. It is an agenda that has been usefully compared to a universalist creed, or Gospel of Competition, promulgated by an alliance of international institutions (e.g. World Bank, World Trade Organisation, International Monetary Fund), global conglomerates and government. Its economic and social tenets; trickle-down, austerity and competition, are held to be beyond dispute, leaving a hopeless exclusion for all those deemed unorthodox or unworthy to enter the inner sanctums of global decision-making (Petrella 1997; Group of Lisbon 1995; George and Sabelli 1994).

Given these origins, it is at best ironic that the Agenda 21 document emphasises environmental protection and social responsibility as primary development aims. It is, after all, the above-mentioned global development priorities which have engendered and presided over the systematic worsening of the common good in both ecological and human terms. As the main geopolitical organisation on the planet, the UN's status as the capable sponsor of a more benign and environmentally conscious world order is questionable. A year before the Rio Summit, the UN agreed to a military operation which took no account of human and environmental impact (i.e. the US deployment of 350 tons of depleted uranium-tipped weapons) and authorised a regime of sanctions against the people of Iraq, unprecedented in its severity (see Clark *et al.*, FAO Report 1996).

Contradiction between the values made explicit in Agenda 21 and the priorities put forward by the global orthodoxy surfaces in praxis as soon as we begin, responding perhaps to the Agenda 21 proposal, to connect the specific experience and conditions of localities, which even in the UK invariably register a perception of things going from bad to worse coupled with strong concerns about the future (see Jacobs 1996), to this international global context. It appears again, in a different way, when we try to measure our ecological footprints; what we do locally against basic environmental sustainability principles like those highlighted in the introductory quotations. Including, as we must in our definitions of sustainability, our social/cultural footprints, the damage and dislocation inflicted on people and communities by the practices of other communities, landmines, debt, food markets and television, damns the contradictions and forces open a credibility gap that is impossible to ignore.

There are, therefore, some serious questions about how, and even whether, the relations between North and South, the so-called developed countries and those deemed to be developing, can be mediated through Agenda 21. For the same reasons, we need answers about the ways in which, and extent to which, any LA21 can grapple with the cause and effect realities of this international global context and connect them with what is happening in communities at a local level. The gap between rich and poor, powerful and powerless and the impact of the one on the other, both between and within nations, remains a defining question for humanity; notwithstanding the signatures at Rio, this gap remains wider than ever.

A21: An agenda for human beings or an agenda for corporate globalisation?

The growth of corporate influence over the last ten years over all aspects of global relations has led to the realisation that the model and process of development being advertised the world over is certainly no longer, if indeed it ever was, government led (Carmen 1997). Under the influence of the rising global orthodoxy, propelled by 1980s monetarist doctrines, governments have, step by step, abdicated their self-appointed roles as prime agents of economic and social development. In their place, unaccountable corporate conglomerates were enabled to pursue an agenda of profit maximization at any cost. The latest example of this process, the *Multilateral Agreement on Investment* (MAI), aims to open all sectors of countries' economies to foreign investment (i.e. large multinational corporations). Unrestrained and free of responsibilities, under the guise of creating new wealth that would somehow percolate to the billions of dispossessed and excluded people of the world, they are free to roam the markets of liberalised finance regardless of social and environmental consequence (see Korten 1997a).

This has produced a situation where the aims of Agenda 21 are supposed to become operational in that same world where the following facts now apply: of the largest 100 economies, half are now corporations; the ten biggest companies have a turnover greater than that of the smallest 100 countries; two-thirds of world trade is conducted by just 500 companies, 40 per cent of it amongst themselves; for every person in the world living on less than a dollar a day, $1,000 are exchanged on international money markets; ten corporations control virtually all aspects of the world's food chain (see Duchrow 1995; Jacobs 1996; Korten 1997a; Petrella 1997; Vidal 1997).

According to UN figures, illegal trade alone is worth $750 billion a year, more than twice the global direct investment of multinational companies in 1995. This has to be set against the backdrop of an out-of-date system which exempts all global transactions (legal or illegal) from taxation. A modest 1 per cent tax would yield $2.50 trillion a year (Swift 1997). A mere fraction of this revenue, according to Petrella (1997), would suffice to provide the three billion humans with access to clean drinking water and the two billion people, who at present have neither, with a shelter. It could also feed the one billion who go hungry.

Such statistics throw into relief the contemporary process that has been dubbed *globalisation*. It is driven by the same global conglomerates which dominate the world's production and distribution of primary resources and materials, basic consumer products, and financial management and investment. These together 'cover all aspects of material and non material production' (Petrella 1997). Moreover, so as to maintain the momentum of their drive for growth and profits, they excercise an overriding influence over the definition and shaping of regulatory structures of international economic policy and trade through their lobbying of and influence over government decision-making (Hines and Lang 1997).

In real terms there is a huge increase in direct corporate control over the vision, forms and means of global development, with a direct and ever-increasing impact on the economic, cultural and ecological life of the planet. This power is underpinned by the deregulation of the checks and balances that political legislature should provide, with a concomitant decrease in the influence of governments and ordinary people over every aspect of social life.

Vandana Shiva recently described this process in terms of the ten commandments of globalisation and corporate rule. These lay down the laws of behaviour for citizens, governments and corporations to enact the corporate agenda as their highest moral duty. The consequences of this predicament are highlighted by the first commandment for citizens – Thou shalt have no right to livelihoods, to work, to food, to water or to a safe environment, and the fourth commandment for governments – Thou shalt give up all functions to protect your citizens and all obligations and duties required of you by your national constitutions (Shiva 1997).

From North to South and back again: global inequality and the widening credibility gap

We should recall that Agenda 21 emerged from the international political process only as a direct result of the long-term pressure brought to bear on national governments and their international institutions by non-governmental organisations and citizens' movements, associations and networks, North and South, through direct lobbying and their *de facto* development counter-practice. This citizens' agenda was (and remains) twofold in its objectives. First, to bring to the fore the reality of global inequity, environmental degradation and social and political exclusion; in short, the utterly unsustainable nature of global *maldevelopment*, which comprises the double folly of both under- and over-development, the latter being the more serious problem of the two (see Carmen 1997b; Amin 1990). The second and consequent priority is to remind all sectors of government and society that the reversal of this universal maldevelopment is literally a question of life and death, if not for ourselves, then for our children, and to charge them with this responsibility. Where previously this agenda had been almost invariably a taboo topic in polite international policy-making company, the first five years of Agenda 21 have presented an opportunity, challenge and (albeit non-statutory) license to bring it nearer to the power centres of social debate and praxis in their local, national and international spaces.

A recent report by the International Council for Local Environmental Initiatives (ICLEI) has shown that more than 1,800 local governments in 64 countries are actively engaged in LA21. In Cajamarca, Peru, for example, a dramatic decentralisation of provincial government structures has gone hand-in-hand with extensive cross-sectoral action planning to address basic issues facing local people in what ranks as one of the poorest communities in the world, '. . . potable water, sanitation, environmental education, and rural electrification. The process has mobilized more than US $21 million for sustainable development activities since 1993' (ICLEI/UNDPCSD 1997: 13).

Nevertheless, the same ICLEI report also shows that it is primarily high income, Northern countries (with the notable exceptions of the US and Canada whose per capita income and consumption rates rank as the highest in the world) which have initiated '. . . 90 per cent of the identified LA21 planning processes' (ICLEI/UNDPCSD 1997: 4). It suggests this may have happened for several reasons; the fact that local government organisations in high income countries were able to participate in the inaugural Earth Summit meeting in 1992; that mechanisms for environmental planning already existing in these countries could be adapted to LA21; and that lack of funding by donor agencies has meant that national campaigns have been slower to get off the ground in middle and low income countries (ICLEI/UNDPCSD: 5).

Notwithstanding the example of Cajamarca, Peruvians remain dispossessed, poor and oppressed, to a great extent on account of the expropriation of material and cultural resources that is the legacy of colonialism and the enormous economic burden they carry on behalf of the cycles of indebtedness and austerity of the last two deacades. Add to this their government's allegiance to the development ideals of the same global orthodoxy that, in its time off from the real business of promoting universal economic growth, attends UN conferences on the environment and development. The European debt to Latin America has recently been calculated as equivalent to the 185,000 kilos of gold and 16 million kilos of silver logged in the Spanish 'Archivo de Indias', plus compound interest (at a mere 10 per cent compared with Peru's typical 20–30 per cent) over 300 years, equalling a figure needing more than 300 digits! (Cuautemoc 1997). The figures, along with the professions of faith and intent, of course, do not add up, and it is in the light of these kinds of perspectives that the credibility gap between North and South, rich and poor, powerful and disempowered, widens.

In spite of statements acknowledging the developed world's causal relationship to maldevelopment, and the necessity of mediating relations between North and South, Agenda 21 has, from its outset, been viewed with scepticism and caution (see, for instance, Middleton *et al.* 1994). In an early response entitled, Waiting for UNCED: *Waiting for Godot*, Johan Galtung (1993) pointed to the normalising effect of reports born of processes like the United Nations Conference on Environment and Development (UNCED); such publications offer the semblance of dealing with problems whilst actually deflecting real concerns into a business as usual scenario (Galtung 1993). This perspective on Agenda 21 has a strong resonance with that recorded by Southern peoples and countries. Writing a response to the original 1992 Rio Earth summit from the point of view of the world's indigenous communities, Jose Dualok Rojas (1994: 50) observed that UNCED 'seemed to function like a commercial market . . . [which is] . . . equivalent to saying that human beings from rich countries can live well while those of poor countries can die'.

His forewarned comparison has been borne out by recent work linking the agendas of lobby organisations allied to transnational corporations such as the World Business Council on sustainable development (WBCSD) and the emergence of a UN consensus around the inevitability and desirability of creating a universal consumer culture within a globalised economy as the only viable route to sustainability. This misappropriation of the concerns, meanings and struggles that brought governments to the Rio summit has become a feature of the rites of the global orthodoxy, and continued throughout the UNCED process and its Commission on sustainable development's monitoring and review up to and during the Earth Summit II conference in New York in June 1997 and beyond that to the Kyoto summit on climate change in December 1997 (see Finger and Kilcoyne 1997; Korten 1997b; Beder *et al.* 1997). Rojas' description of the Rio proceedings concluded sardonically: 'Was it good or was it bad? Only time will tell' (Rojas 1994: 50).

What time has already told us is that, notwithstanding examples of good practice and pioneering programmes from around the world, governments by and large have been unwilling to deliver, or even address, even a few of the targets they identified and committed themselves to at Rio. Whilst the resources and practical programmes to achieve these targets have been found massively wanting, the corporate lobby's aim to win international consensus on public policy and subsidies for market incentives has fast become the evangelical discourse of the 1990s. A recent brochure on Pakistan aims to reassure us of the country's development priorities. The publication surveys the performance and incentives of key industrial sectors, affirms its commitment to economic growth and complete privatisation, and carries the subtitle 'Business as usual – open for investment' (Images, Words Ltd 1996).

And so a country's future development is entirely predicated on the tenets of the global orthodoxy. There is no mention or concern that less than 2 per cent of GDP goes to education in a country where nearly one in three people is illiterate (Khan 1996). No question of the appropriacy of exporting an entirely alien economic model based on competition and the moral imperative of profit to Pakistan, a society whose relational, community-based economies, are millennia-old and firmly embedded in a culture of reciprocity and mutual responsibility (see Cultures & Development 1995 and Brahimi 1996). No inkling of the way in which such cultural and economic aggressivity creates enormous socio-psychological tension and fuels opposite and extreme reactions such as religious fundamentalism in the target cultures (el Saadawi 1998). There is even a shadow of a doubt about the way of life that is first, universally impossible by the same economic laws that are presented as its guarantee and, second, a perilous fantasy when measured against the realities of the earth's resources and the actual environmental and social consequences of such a model. We have our own societies that have already elaborated this model to testify to that, with the mega-city becoming inhabited only by 'those who don't eat' and 'those who don't sleep' (de Morais 1997). This literal description also traces the demarcations between North and South, the patterns of social exclusion within and across societies. It hints at the human consequence of maintaining such divisions. Those who do not sleep seek protection of their life style from the rest of excluded humanity behind steel-bars, locked gates and fortifications. They do not sleep for fear of those who do not eat.

Sustainability and the global democratic deficit: the citizens' agenda

> The approaching millennium promises to accelerate the processes of globalization which are already irreversible and advancing dizzily before our eyes to envelop even the most remote people of the world in a suffocating embrace. Structures, values and ideas that we took for granted for most of this century are disintegrating. Global answers are required if we are to adjust to and understand one another.
>
> (Ahmed 1997: 1)

If the entire population of the earth currently lived as the average North American or Canadian we would already require three planets of the proportions and ecological characteristics of our own to sustain them (Wackernagel and Rees 1996). This simple sustainability equation does not even begin to deal with variables such as the continuous degradation of natural resources, the effects of environmental pollution and the social and cultural consequences of economic deprivation and inequalities. The major challenge of sustainability, as Ahmed suggests, is to find ways of developing relationships between North and South, the richer developed and poorer developing worlds, that can replace those offered by the global orthodoxy. Where exploitation, competition and ignorance dominate, we must sow the seeds of collaboration, co-operation and cross-cultural understanding.

The tensions between the proposal put forward by Agenda 21 and the realities of global maldevelopment are perhaps nowhere more evident than in its apparent faith in government as an agent capable of delivering ready-made solutions for bringing about benign and lasting change. Given the context of the corporate agenda for globalisation within which Agenda 21 appeared, such faith appears ill founded. Truly sustainable development cannot be planned as a grand scheme and imposed on people, rather it must be endogenous, arising from people's own specific experience and situations. It does not share the one-dimensional model of top-down/bottom-up that informs the language and thinking of Agenda 21, but works rather from the 'inside out', where an autonomous citizen's agency initiates a dialogue process with institutions and expertise, as and when it identifies such a need for itself. If government is to have a role in this at all, it is only to support, facilitate and co-ordinate the process in a fully transparent and accountable way (Rahman 1993; Carmen 1996).

In describing the international dimension of maldevelopment, Susan George suggests a model of two spheres – North and South – where the core, controlling elite (the hierarchy of the global othodoxy) of the North interacts directly with its core Southern counterpart. Relations with the periphery (i.e. the people and the citizens) of each sphere are mediated and governed only through their respective cores. According to this model, institutional power in the North directs or coerces the people in their development and, in the case of the South, people and governments alike. George suggests that the alternative to this closed circuit of power is to open up genuine relations between the peripheries of both spheres, i.e. between the people and citizens of the South and those of the North (George 1991).

This approach of developing links and communication between people, their communities and organisations in different parts of the world is the first crucial step towards addressing the central question of *governance* over international development priorities. It is clearly within the scope of government, whether local, regional, national or supra-national, to facilitate and promote such initiatives. Indeed, Agenda 21 documents have encouraged and emphasised such approaches to addressing North–South relations, and the impulse that gives rise to them pre-dates Agenda 21 (Local Government Management Board 1995). The seeds from which such approaches may emerge have been nurtured by the global development education movement and its pioneering of practical linkage between community organisations and local institutions. Recent reviews of these programmes in the UK and Europe have pointed to their success in combating the cultural fall-out of prejudice and misunderstanding that is the residue left behind by the historical continuum of colonialism, development and globalisation in the North, as well as to their educational value in realising the true connectedness of people and their actual effects upon each other's lives (Gerrard 1995).

However, it is significant that the same reserve with which Southern people have met Agenda 21 is present in many of their responses to such linking initiatives. One participant in a programme evaluation states that Northern communities need 'to listen and respond, rather than hearing and doing nothing' The need to initiate action for change in the North is clearly a crucial step forward in focusing links into coherent responses to the context of globalisation, and LA21 and the debate over sustainability and global futures has provided a basic framework within which this imperative may be addressed and interpreted. For example, links between the Gulu region in Uganda and the County of Lancashire in the UK have been made explicitly around sustainable development and LA21. Independent NGOs are being established at each end of the link to facilitate a joint programme of sustainable development projects, steered by representatives of the respective communities. Relations between the Borough of Stevenage and Kaduma in Zimbabwe have been developing since 1989 and, following the Rio summit, the link has addressed the issue of development and North–South relations by highlighting the learning potential across the two cultures, in particular by emphasising what Stevenage can learn from Kaduma (Thorpe 1997). Gloucestershire's NGO and community-led Vision 21 process organised a national conference where its own link communities came together with participants from all over the UK to explore issues of sustainability and global inter-dependence (Vision 21 1996).

In Luton, UK, collaboration between local authorities and the World Wide Fund for Nature UK (WWF UK) has enabled the piloting of a joint initiative with Peshawar, Pakistan, for linking and learning about sustainability through local and international partnerships between community groups, schools and their local authorities. The project emphasises the equality of the learning partnership that is being developed between both places. The contact with, and example set by, organised women's groups in Peshawar, for instance, in addressing environmental and health issues, has encouraged women in Luton to address these issues in their own communities through learning about local organic food growing. The programme also aims to increase people's involvement in decision-making about local development and to link up community groups in both places to share experience and understanding on local and global issues. Above all, it makes explicit and tries to facilitate the parts played by local communities and their authorities in global development, and promotes the importance of, and responsibility for, cultural and practical change in the North as much as in the South (Lubelski 1996).

Such practical responses to the imperative for change for the better in North–South relations, the bulk of the burden of responsibility for which falls on the North, indicate the potential for 'opening up international development to society as whole', and some first steps towards 'building global civic society' (Fowler 1996: 12). By following through this potential, we may come nearer to making up what amounts to an enormous global democratic deficit, where decision-making over the vision and forms of global development is the sole preserve of powerful elites, organisations and institutions (Korten 1990, 1997a, 1997b). A different kind of global citizenship is indicated by such approaches to international relations, one which emphasises collective rights and responsibilities across cultural and international boundaries. It is one which is in tune with growing calls being made (beyond and within LA21 processes) for a 'broadening and deepening of all forms of democracy in all areas of life'. (Burns *et al.* 1994: 179). Significantly, the need of poor and culturally excluded groups in the North to resist the globalisation agenda is already being registered at the level of local government institutions 'the impact of globalisation and remoteness . . . (induce) . . . a perception that solutions at a local level are perhaps no longer feasible or possible' (Jalal Uddin 1997: 6).

Telling the whole story: learning from the South and the agenda for change

The marginalisation experienced by the communities which Jalal Uddin describes, which is common to excluded communities the world over, is of course continually resisted in diverse and creative ways. What is most relevant for this discussion is the origin and location of this creative resistance in the South.

The notion that Southern peoples might in fact have more to offer the North than the North has to offer the South, in terms of practical know-how in working towards sustainable development, is relatively new, at least in mainstream terms. However, organisations such as CAFOD and Oxfam are now actively promoting Southern over Northern experience and the kinds of reciprocal learning that the more forward-looking LA21 programmes aim to bring about. CAFOD recently brought organisers from the Comedores movement in Lima, Peru to Lithuania in order to share experience of finding solutions to human suffering in the midst of collapsing economies through setting up communal kitchens and the educational activities that surround them (Jacobs 1996). Oxfam's decision in 1995 to start tackling social inequality in the UK was born of the realisation that, first, the structures of inequality and the causes of exclusion are 'not just international but within communities and countries everywhere' and, second, the partnerships it had built up in the South had released the creative vision and means by which this structure could be addressed in the UK. 'By learning from others, we multiply the possibilities of overcoming problems with which others are already struggling' (Roche 1996). Grassroots networks such as the Spanish-based RICA (International Network of Alternative Communication), set up to share experience of resistance to marginalisation, exclusion and the globalisation process, have also emphasised the commonality of problems and issues facing people South and North, and the great potential for discovering and learning solutions through direct contact and communication (Zobel 1997).

For those who do not consume, the hundreds of millions of displaced and dispossessed urban dwellers, the long-term unemployed, surviving as best they can by means legal and illegal, the overbearing problem is not consumption but *production*. As was briefly discussed above, our present world, not least the bulging Third World megatowns such as Mexico City or Karachi, can be broadly divided between those who do not eat and those who do not sleep, where only the social elites and mega-rich are able to afford themselves protection behind fortifications from the desperate criminality being forced upon their fellow citizens. It is clear, however, that those who do not eat are in that situation because they do not produce, and are systematically excluded from the official productive processes of society. They represent the capitalism of poverty, those who are not needed any more (even to be exploited), the 'off-shore islands in a sea of plenty' (Latouche 1993: 154). Working in those same globally excluded Southern communities, Third World activist and academic, Clodomir Santos de Morais has pioneered a proven method of overcoming this exclusion over the last 30 years in Latin America, the Caribbean, Africa and Europe. His method of recreating autonomous local economic networks through massive organisational workshops (OWs) contains many lessons for the welfare-dependent deprived areas of the North, too. For the excluded to eat they have first to be capable of producing, and this the OWs do massively, creating worker-owned, democratically self-managed production and service enterprises (De Morais 1997).

For local communities, North and South, to withstand and outlive the pressures and instabilities of the globalised market economy, we need solutions which can relocalise their economies in terms of resources, needs and people. One of the greatest impacts that the consumerist practices of Northern communities have on those of the South is that of the global food system, which is subject to ever-increasing concentration in the hands of corporate monopolies. This, combined with the super-industrialisation of agriculture, leaves whole communities vulnerable and exposed to the manipulation of international markets whilst their environments become degraded and ever more contaminated (Trainer 1995). At the same time, excluded people in the North have long been subject to nutritional poverty due to lack of access to affordable fresh and unprocessed food, whilst their environments are blighted by the same consequences of the development of industrial agriculture and the absurd system of national and international transportation that governs food distribution.

The potential for the reduction of Northern footprints, for a real change from destructive, insane consumption to sustainable, healthy production, is nowhere better illustrated in the North than in the move to reinvigorate local food economies, particularly through organic urban food growing and community owned food co-operatives (Davis *et al.* 1997; Egerton 1998). These strategies are, in effect, direct globalisation counter-practice. It is no surprise, then, that they draw on the techniques and patterns of traditional food economies and the huge body of experience of urban agriculture in the South. One such example, the Arid Lands Initiative in Salford UK, which has created an organic food system linked into a community café around a no-hope 1960s tower block, was inspired by the urban forest gardening techniques of the Yemen in the south of Arabia (LGMB 1997).

Conclusion: whose agenda for the twenty-first century?

At the start of the 1990s, David Korten, an experienced practitioner and commentator at the forefront of the debates about international development, issued a call in his book *Getting to the 21st Century* for people across the world 'to join hands in the spirit of global citizenship to define and implement an agenda for social transformation' (Korten 1990: 61). The international dimension to Agenda 21, the real agenda of globalisation and development, is the actual context of all present-day social, cultural and intellectual work. It therefore presents some important challenges to those involved in striving for a more sustainable future in the UK and beyond. Stated plainly, and taking a view inclusive of Southern and indigenous peoples, it is a question of whether or not we face up to, come to terms with, and address globalisation explicitly as part of LA21 programmes. For local authorities, this means a redefinition of their role as conscious contributors to a just and sustainable international order. Issues such as human rights and equity for all, community-based decision-making over local development, the promotion and design of alternatives to the culture of consumption, and the development of self-reliant and ethically conscious local economies, need to become the life-blood of local and national government policy making and service provision. This implies that local authorities and organisations dedicated to public service are *either* in the business of supporting an agenda for change owned by people in a spirit of international co-operation *or* remain party to a palliative exercise that appears to be designed to mask the illusory agenda of a universalist consumer culture regardless of the human and environmental cost.

References

Ahmed, A. (1997) *Jinnah, Pakistan & Islamic Identity – The Search for Saladin*, London: Routledge.

Amin, S. (1990) *Maldevelopment: Anatomy of Global Failure*, London: Zed Books.

Beder, S., Brown, P. and Vidal, J. (1997) 'Who killed Kyoto?' *The Guardian*, Wednesday 20 October.

Brimah, N. P. (1997) 'Structural adjustment in Ghana', unpublished correspondence with the authors.

Burns, D., Hambleton, R. and Hoggett, P. (1994) *The Politics of Decentralisation – Revitalising Local Democracy*, Basingstoke: Macmillan.

Carmen, R. (1996) *Autonomous Development – Humanizing the Landscape*, London: Zed Books.

Carmen, R. (1997a) 'In the wake of Homo Oeconomicus: Homo Mundialis? The countervailing human agency of civil society; a definite beacon of hope', paper presented to 22nd SID World Conference – 'Which Globalization: Opening Spaces for Civic Engagement' Spain, July, http: \\ redtips.org\ tips\ forum\ sid\ temaz.htm.

Carmen, R. (1997b) 'Producer works!' in *Development* 41: 1 (forthcoming).

Clark, R. *et al.* FAO Report (1996) *The Children are Dying – The Impacts of Sanctions on Iraq*, New York, World View Forum Inc.

Cuautemoc, G. (1997) 'The real foreign debt', in *Resurgence*: 184.

Cultures & Development (1995) 'Entrepreneurship and African cultures', Synthesis of Papers and Debates, Brussels.

Davis, L., Middleton, J. and Simpson, S. (1997) 'Breaking new ground: initiatives towards community agriculture in the Metropolitan Borough of Sandwell, England', paper presented to the International Conference on Sustainable Urban Food Systems, Toronto.

De Morais, C.S. (1997) 'Efficient labour markets: a challenge for Latin America', keynote speech at the K. Adenauer Foundation International Seminar, San Jose, Costa Rica.

Duchrow, U. (1995) *Alternatives to Global Capitalism*, The Netherlands: International Books.

Egerton, L. (1998) *Local Food for Local People: A Guide to Local Food Links*, Bristol: Soil Association.

el Saadawi, Nawal, (1998) 'All I said was don't bomb Saddam . . .', *The Observer Review*, 22 February.

Finger, M. and Kilcoyne, J. (1997) 'Why transnational corporations are organizing to save the global environment', *The Ecologist* 27:4.

Fowler, A. (1996) 'Authentic NGDO partnerships in the new policy agenda for international aid: dead end or light ahead?' in *Development and Change*, June.

Galeano, E. (1992) 'Visions of Tomorrow – South', Channel 4 television, 12 June 1992.

Galtung, J. (1993) 'Waiting for UNCED: Waiting for Godot', in Rajan, V. (ed.) *Rebuilding Communities*, Totnes: Green Books/WWF.

George, S. (1991) *How the Other Half Dies*, Harmondsworth: Penguin.

George, S. and Sabelli, F. (1994) *Faith and Credit – The World Bank's Secular Empire*, Harmondsworth: Penguin.

Gerrard, P. (1995) *Grassroots Development Education – Guide to Good Practice*, The Hague: Towns and Development.

Group of Lisbon (1995) *Limits to Competition*, Cambridge, MA.: MIT.

Hines, C. and Lang, T. (1997) *The New Protectionism*, London: Earthscan.

ICLEI/UNDPCSD (1997) 'Local Agenda 21 Survey – A Study of Responses by Local

Authorities and their National and International Associations to Agenda 21', New York/Toronto.

Images, Words Ltd. (1996) 'Pakistan: business as usual – open for investment', *The Observer*, 29 December.

Jacobs, M. (1996) *The Politics of the Real World*, London: Earthscan.

Jalal Uddin, R. (1997) 'Community Action' in *Asian Times*, 2 September.

Khan, I. (1996) 'National education movement in Pakistan' in *Islamica* 2.3., ISLSE.

Korten, D. (1990a) *Getting into the 21st Century*, Kumarian Press.

Korten, D. (1990b) 'NGO Strategic Networks: From Community Projects to Global Transformation', PCDF, Internet Publication.

Korten, D. (1997a) *When Corporations Rule the World*, London: Earthscan.

Korten, D. (1997b) 'The UN and the Corporate Agenda' People-Centred Development Forum (PCDF) Internet Publication.

Latouche, S. (1993) *In the Wake of the Affluent Society*, London: Zed Books.

Local Government Management Board (1994a) *Sustainability Indicators Report*, Luton: LGBM.

Local Government Management Board (1994b) *North/South Linking for sustainable development*, Luton: LGMB.

Local Government Management Board, UK (1997) *Local Agenda 21 Case Studies – Case Studies of Work for Local Sustainability*, London: LGMB.

Lubelski, M. (1996) 'LAPIS (Luton And Peshawar Initiative for Sustainability) – Local Agenda 21 and The International Dimension', paper presented to the European Conference on Sustainable Cities and Towns, Lisbon.

Middleton, N., O'Keefe, P. and Moyo, S. (1994) *The Tears of the Crocodile: From Rio to Reality in the Developing World*, London: Pluto Press.

Petrella, R. (1995) 'Politics has abdicated in favour of Private Enterprise', interview in *The Courier*, EU, Brussels, No.115, May–June.

Petrella, R. (1997) 'On Globalization', unpublished talk given to the Conference on Globalization and the Vitality of Cultures, Réseau Cultures, Brussels.

Rahman, M. D. Anisur, (1993) *People's Self-Development*, London: Zed Books.

Roche, C. (1996) 'Southern comfort', *The Guardian*, 20 November.

Rojas, J. D. (1994) 'UNCED: Ethics and development from the indigenous point of view' in Brown, N. and Quiblier, P. (eds) *Ethics & Agenda 21*, New York, UNEP.

Shiva, V. (1989) *Staying Alive – Women, Ecology and Development*, London: Zed Books.

Shiva, V. (1997) 'The Ten Commandments of Globalisation and Corporate Rule', talk given to World Development Movement Meeting, Manchester, UK, 28 June.

Swift, R. (1997) 'Just *Do* It', *New Internationalist*, November.

Thorpe, E. (1997) *Links as Development Education Resources – The experience of LADER Project at the Lancashire Development Education Group, Lancashire*: LDEG.

Trainer, T. (1995) *The Conserver Society – Alternatives for Sustainability*, London: Zed Books.

Vidal, J. (1997) *McLibel – Burger Culture on Trial*, Basingstoke: Macmillan.

Vision 21, (1996) *The Global Footprint – Symposium Report*, Rendezvous: Gloucestershire.

Wackernagel, M. and Rees, W. (1996) *Our Ecological Footprint – Reducing Human Impact on the Earth*, Canada: New Society Publishers.

Zobel, G. (1997) 'Horsemen of the Apocalypse', *The Guardian*, 13 August.

10 Equal opportunities and local initiatives in Sri Lanka

Anoja Wickramasinghe

The preparation of an environmental agenda for Sri Lanka for the twenty-first century has been extremely slow. This is due to the complex nature of the task, poor communication between grass-roots and state agencies – both between people and decision makers and among sectors. Many have realised that the adoption of sustainable development and the basic principles of Agenda 21 requires policy adjustments and an integrated approach. The present situation shows that inequalities in development opportunities and resource distribution have resulted in depriving local people of decision making in the management of the local environment. Locally acceptable appropriate technologies have been withdrawn with the introduction of externally developed ones. The question in hand is not simply related to the adoption of Agenda 21 alone, but is a matter of restoring local initiatives. Resources and benefits from development are unequally distributed and, as a result, the scope in which the deprived sectors may act is marginal.

This chapter will examine what has been done so far in the preparation of Local Agenda 21 for Sri Lanka, and whether or not attempts are being made to empower local communities and initiate from local innovations and culture. I will also discuss whether the overriding dominance of perceptions and attitudes towards 'dependency' and 'beneficiary' are being eroded in order to create equal opportunities.

Agenda 21 as external pressure for action

The development agenda for the twenty-first century and also for the remaining few years of the twentieth century is difficult to grasp and focus upon. In order to achieve the goals of sustainability, on top of the internal unrest and the problems pertaining to meeting the needs of the present generation, Sri Lanka should develop long-term policy frameworks to bring about a practical integration of the basic principles of Agenda 21; which includes a holistic approach. Sri Lanka is far behind in terms of formulating its LA21. The situation is similar regarding the local level policy framework for sustainable development. The adoption of the concepts of equal opportunities and local initiatives as people's innovations, is rather difficult due to well-established socio-economic disparities and the bureaucratic administrative structure. The promotion of the disadvantaged, particularly rural people including women and indigenous groups, to the forefront of policy formulation and implementation is a challenge. This is mainly due to the need to reverse the current process of making decisions by the policy makers and the affluent on behalf of the poor, who have until now been considered 'recipients' and 'beneficiaries'. Until power is shared between these two groups, Agenda 21 will not be accepted.

Sri Lanka's commitment to enhance national efforts to promote sustainable and environmentally sound development strategies has been accepted politically. The broad set of policies that should be adopted in favour of Agenda 21 is comprehensive and connected with all the development sectors that are now being handled by a well-established sectoral framework. Since the United Nations conference on Human Environment held in Stockholm in 1972, the national concerns regarding environmental issues have increased. Similarly, the 1992 Convention on Biological Diversity has encouraged a more conservation focused strategy, to restore and conserve biological resources. Moreover, 'A strategy for sustainable living' presented by the World Conservation Union (IUCN), the United Nations Environment Programme (UNEP) and the World Wide Fund for Nature (WWF) in 'caring for the earth', shows fundamental elements pertaining to conservation and development. The major features of this strategy include the establishment of links between conservation and development, and acknowledging the importance of enabling communities to care for their own environments.

The externally driven demand for local action has been strong, in the sense that much has been adopted through commitments to international conventions and ratification rather than as self-perceived problems to which the country is prepared to find solutions. The more serious problem is the lack of financial strength in the long run to proceed along proper lines and concentrate on local concerns. Unless the principles are connected with those who deal with the local resources in their day-to-day activities, and who have the control over the most crucial resources such as land, forest and water, and establish feasible and locally appropriate mechanisms, the process will not be internalised. In the phase of mitigating and reducing the impacts of global environmental problems pertaining to climatic change, ozone layer depletion, and air and water pollution, that have already emerged and threatened human survival, local action agendas need to be promoted. The country's formally endorsed global partnership has a number of practical implications. One important aspect is that it has created a political commitment to examine the direction of development. This has, at least in academic circles, created greater pressure to investigate the environmental issues connected with development. Although many of the ideas are forcefully embedded, the relevance of the proposals in Agenda 21 can be justified by referring to the degradation and depletion of resources, their socio-economic implications, and the unsustainability of development initiatives introduced during the last few decades.

Despite investments made to develop Sri Lanka, crucial problems like deepening poverty, reductions in land productivity, land degradation, deforestation, siltation and sedimentation of reservoirs and drainage, crop damages, floods and landslides, are on the increase. In the interest of achieving target economic goals, the country has not made adequate efforts to account for the full costs of economic activities, particularly the damage done to the country's natural resources on which the majority of the people live. The people are the victims of interventions: for example, tobacco cultivation expanded into the sloping terrain of the highlands has offered income opportunities for the farmers, and it has contributed a substantial amount of revenue to the state. As a result of four decades of over-exploitation of resources, the company has withdrawn from the area. There is no system to recover the loss of environmental resources and their cost unless the country goes to international lenders to get funds for subsidies for those who have been trapped, and also for the conservation of land which cannot recover for several decades. The present situation reveals that there is a crucial need for restructuring the policy framework to ensure that neither the local resources nor the local resource managers are exploited for short-term revenues. This push for capitalist economic growth and the increasing corporate influence over the development process of many 'developing' countries is explored in more depth by Lubelski and Carmen in Chapter 9. They argue that this destructive process has its roots in 'globalisation' and 'maldevelopment' reflecting international development priorities which do not support environmental protection and social responsibility as the main development objectives.

The action framework

At the local level, the process of formulating a policy framework and policy reforms to achieve the goals of sustainable development has been extremely slow. One reason is that in the compartmentalised and structured administration there has been no responsible agency/sector to undertake the whole task of formulating local Agenda 21. The top level agencies formally responsible for providing state mechanisms for the sustainable management of the country's resources have either limited capacity or prioritised mandates. Agencies are often guided by the sectoral strategies formulated either in isolation or with a focus driven by external agencies. Each sector (including agriculture, plantation, health, housing, coastal conservation, forestry, energy, environment, fisheries, education, industry, irrigation) has a number of planning units. The basic question that arises here is to what extent this existing structural framework would permit the adoption of a holistic approach to hold and reverse environmental degradation and to promote an environmentally sound sustainable development framework for the twenty-first century.

In the history of national efforts pertaining to environment and sustainable development, major environmental issues have been elucidated in Sri Lanka's national report prepared for UNCED. The National Environmental Action Plan (NEAP)(Ministry of Environment and Parliamentary Affairs, 1991) has for instance, identified local environmental issues and proposed possible mitigatory measures. The first comprehensive environmental plan for the period between 1992–96 has been prepared to establish an environmental agenda within the context of resource management and development. It is clear that rather slow attempts have been made to execute the proposed action.

Prior to the preparation of the NEAP, the sectoral plans and programmes pertaining to forestry, agriculture, coastal and marine ecosystems, watershed and mountain ecosystems had been focused on the conservation of natural resources. The Soil Conservation Act of 1951 had made provision to introduce precautionary measures to protect land from erosion, although it had not been enforced as expected. It is not the lack of policies and legal support that obstructs conservation, but the responsible agencies with the capacity to make a change. Inter-sectoral co-ordination is poor due to the very nature of a sectoral focus. As a result some sectoral policies like agricultural policies are not in favour of forestry or biodiversity, nor of community participation and empowerment.

Moreover, the interests of external agencies are specific and leave gaps. International agencies like UNDP help strengthen the institutional capacities to manage the resource base. There are many agencies directly involved primarily on sectoral development. For instance, the Asian Development Bank (ADB) has provided a loan for the projects related to the development of the fisheries sector, land use planning, water resource development and industrial pollution control. The USAID has supported the Natural Resources and Environment Policy Project (NAREPP) – the goal of which is to support the government in the management of natural resources. The Norwegian Agency for International Development (NORAD) also assists the government to promote natural resource management. The Netherlands has extended its assistance to focus on the protection of wetland ecosystems and industrial pollution reduction. The governments of the UK, Canada, Denmark and Finland assist Sri Lanka in promoting environment management in the areas of forestry and waste management, and many other agencies are also involved in providing assistance to a number of other sectors such as infrastructure development, poverty alleviation, human settlements, health and education. In almost all these cases, provision has been made to incorporate environmental considerations, but most agencies suffer from the inability to implement these programmes.

Two Ministries are directly dealing with the issues of environmental and natural resource management and overall development which leads to fragmentation. The first is the Ministry of Transport, Environment and Women's Affairs (MTEWA) whose responsibility it is to formulate policies in respect of the environment and natural resource management. The second is the Department of National Planning (DNP) operating under the Ministry of Finance, Planning, Ethnic Affairs and National Integration which has a clear mandate on Agenda 21. The overwhelming importance of the DNP is that it is responsible for co-ordinating planning across all sectors, between central and local levels and also in approving public sector expenditure for development. Sri Lanka has been fortunate in securing the support of the 'Capacity 21' of the UNDP, which is designed to build up local capacity to integrate the principles that help link environment, economic and social development.

The present situation shows that, in a broader context, an attempt has been made to ensure that sectoral development programmes are congruent with the goals set by Agenda 21. However, the lack of an effective mechanism to co-ordinate the sectoral programme has impeded the interaction, communication and inter-sectoral linkages among various actors who are involved in addressing the environmental issues in a broader, practical, development context. To overcome this limitation the government established the National Development Council (NDC) in 1996, headed by the President. These trends suggest that the primary concerns have been to examine the 'environmental impacts' or environmental implications of the sectoral interventions, and ensure that mitigatory measures have been incorporated. The problem is not related to the lack of policy-level integration, but that things are not happening on the ground. For instance, policy support has been given to lease out the abandoned and degraded tea lands to the private sector for other uses, on the assumption that either environmentally friendly measures or farming practices will be adopted. In reality, more exploitation is taking place due to the felling of trees, cultivation of potatoes and short durational vegetables for the market. On the one hand, the on-site destruction is unrecoverable and on the other these transitions in the upper-river catchments contribute heavily to the siltation of reservoirs. The deepening problems in the Uma Oya catchment and the siltation of Rantambe reservoir demonstrate how the lack of inter-sectoral connections cause serious damage to the nation. This also reflects the lack of linkages between national policy and local grass-roots action. From the perspective of the grass-roots, where the changes are to take place, a clear marginalisation of the people with whom the task needs to be shared can be seen.

I argue that LA21 should be directly based in the local context. It should reflect how future plans intend to reverse the degraded conditions and build up the initiatives of local people, and should show how it intends to provide equal opportunities in economic development and resource management. The policy decisions that facilitate the adoption of appropriate technology for the betterment of the lives of local people, especially women, need to be taken in partnership with local people. The local pressure for adopting the principles set in Agenda 21 is exerted by the people who have withdrawn their partnership due to the top-down and externally driven development process that has demoralised their spirit.

Inequitable living standards

The provision of opportunities to improve living standards is crucial for the majority of the population in Sri Lanka. Inequalities in the availability, access and rights to basic needs such as food, shelter, water, wood-fuel and services are well marked among social sectors as well as geographical areas. Not only does one-half of the population live below the poverty line, and depend on state delivered free food subsidies, but they also suffer from poor quality and scarcity of resources on which they depend to meet basic needs. Moreover, the benefits of the assets allocated to improve living standards are unequally shared. As much as 90 per cent of farmers are Janasaviya (poverty alleviation support) recipients (Sri Lankan

Environmental Journalists Forum, 1994). Environmental resources conservation, livelihood maintenance and development are inter-connected. Of a total of about 19 million, nearly 79 per cent of the country's population live in rural areas while about 48 per cent directly depend on agriculture for income and employment. This implies that for the majority of the population the biophysical resources are of immense importance. When land ownership is taken into consideration, it is clear that nearly 80 per cent of the land, including all common property resources that are considered as part of the survival systems, is owned by the state. The people, particularly the ones who have neither economic options nor adequate resources, suffer from hardships, and have become dependents on state deliveries. The urban population amounts to about 21 per cent with a growth rate of about 1.2 per cent a year (lower than the national population growth rate of 1.5 per cent). Many disparities exist in regard to the distribution of basic requirements. For about 85 per cent of the rural population pipe-borne drinking water is a luxury, and this proportion depends on wells. Another 10 per cent use surface water sources like rivers and streams – which implies that only 5 per cent have access to pipe-borne water.

The Environmental Awareness Survey (1992) showed that the environmental impact on health is the major environmental issue in Sri Lanka. The problem of safe drinking water is a priority issue of environment, health and rural development. The SLEJF (1994) report that poor water supply and excreta disposal systems have resulted in 40 per cent of the Sri Lankan population being affected by typhoid, amoebic and bacillary dysentery, infectious hepatitis, gastroenteritis, colitis and worm infections. The situation with regard to energy for domestic cooking is also crucial, since about 92 per cent of households depend on fuelwood for domestic cooking and it is the sole source of energy in rural areas. Kerosene is the source of energy for lighting for about 82 per cent of the housing units, and in the rural areas only 8.3 per cent of houses have electricity for lighting (NARESA, 1991).

The contributions made to improve the quality of life under these circumstances have been small, and as such, an action agenda is urgently required to mitigate the negative environmental impacts on the lives of those who directly depend on environmental resources. Each of the six themes of Agenda 21 that set out the basis of substantive action programmes to foster the sustainable use of natural resources for human development, while ensuring equitable living standards and equality of life in a clean and sustaining environment, is relevant to the local context. All these themes – Revitalization of Growth with Sustainability; Sustainable Living for All; Human Settlements Development; Efficient Resource Use; Global and Regional Resources; and Managing Chemicals and Waste – require people's responsible participation. UNCED stated in *A Guide to Agenda 21* that:

> An essential ingredient for success and an early realization of these goals is the active and full participation of all relevant groups, including women, youth, indigenous people and their communities, non-governmental organizations, farmers, local authorities, trade unions, business and industry and

> the science and technology community. The diverse backgrounds, skills and experiences these groups can offer are essential to the transition to sustainable development.
>
> (1992: 14)

Despite the state expecting and preaching economic growth, for the majority this has not satisfied their expectations, although they bear the burden of environmental problems and their impacts. The reduction of agricultural productivity, especially plantation crop production in the highlands, is turning into a national crisis. Nearly 350 tons of surface soil is lost per hectare annually from the lands that are under seasonal vegetables and tobacco. These issues pertaining to development are not merely environmental issues but include matters related to inequitable control of resources. Over the decades of intervention, development has been separated from the people, and achievements have been shared more by those who have the capacity to reap the benefits than by those with below average living standards. These achievements include the benefits of hydro-electricity projects, river valley diversion, water supply services, and top-level employment opportunities created in association with trade liberalisation and education.

One fundamental aspect that has not been taken into consideration in the process of accompanying the principles of Agenda 21 is resource ownership. The role of land-based resources in providing basic needs as well as the living standards of the people of an agrarian economy cannot be overlooked. The culture related to the sharing of resources, water bodies, forests, common land and riverine vegetation has been rejected in introducing state ownership over resources. This implies that people have become unauthorised users, and they use resources on customary grounds so cannot make decisions regarding improvements. Modern technological interventions have therefore done greater damage by way of removing the self-reliance of those who have managed the resources indigenously. The propaganda in favour of modern technology displaces the locally evolved technology and replaces it with an externally developed, prescribed one that is costly and gives short-term benefits. The recent trends in the economy have created more problems for the grass-roots – women, small-scale farmers and indigenous communities in particular.

Tools such as Environmental Impact Assessment (EIA) and environmental monitoring and evaluation have been adopted to ensure that environmental issues are taken into consideration. However, such tools will not uplift living standards. In the absence of primary conditions such as equal rights to resources, subsistence and basic needs, popular participation in environmental restoration and in linking environment concerns and development is impossible. If opportunities are for specific social sectors, especially for those who have a keen interest in control over poverty stricken social sectors, then the environmental goals will only be technically integrated at policy levels. In addition, the development process should establish mechanisms for an equitable living standard for the present and the future. As has been noted (Beltratti *et al.*, 1995; Solow, 1974), inter-generational equity is an essential element in this process, and from the perspective of both environment and sustainable development the attention paid to the inter-generational perspective of development has been insignificant. For instance, the problems of land degradation (e.g. sedimentation of newly constructed reservoirs resulting in reduced capacity to generate hydro-power and feed irrigation; deforestation that exposes about 40,000 hectares of land annually) are the result of seeking short-term benefits.

The gap between the better-off and the poor, policy makers and the grass-roots, as well as gender inequalities prevents many of the principles of Agenda 21, such as equal participation, being implemented. Agenda 21 in these circumstances does not mean simply the policy integration of environmental concerns to development, but measures to eradicate inequalities. There is much criticism of some principles of Agenda 21 with regard to their inadequacy to foster equality. While Reed (1996) for example, argues that these principles contradict specific goals such as economic development and capacity building through scientific and technical knowledge, Shiva (1993) argues that the 'mainstream environmentalists' – as manifest at the Earth Summit – are divorced from feminism and perpetuate capitalist patriarchy.

However, there are no simple answers to local problems. The wide spectrum of situations, disparities, injustices and inequalities needs to be taken into consideration to formulate a locally adoptable national agenda. The action that has been taken to empower and incorporate the concerns of those who have been marginalised in the past, in the process of development, can be explained only by examining the realities in which people act, behave and have been responded to.

Equal opportunities and popular participation

From the level of diagnosing local problems and identifying development needs, to the level of executing, monitoring and evaluation, a dominant role has been undertaken by the decision makers, executing agencies, and in some cases by the donors. The externally driven models of modern development have displaced local initiatives, and made local people withdraw. Studies conducted in the dry zone reveal that in the villages in Kelegama, Muriyakadawala, every effort has been made by the extension services to promote crop production for the market. Such efforts have resulted in excluding the locally appropriate environmentally sound resource management practices. Market orientated systems have been introduced – with external input, subsidised crop varieties, inorganic fertiliser, and chemicals for weed and pest control. In the intermediate zone of the highlands, tobacco – a market-guaranteed crop – was established about 40 years ago with subsidies delivered to the farmers. Since the late 1980s, this assistance ceased and the Ceylon Tobacco Company has withdrawn from the area due to the decreasing return on its investments. In these circumstances, the local people have been used to facilitate the process of executing the externally driven profit-orientated projects. The 'target groups' or 'the beneficiaries' of development – according to the conventional development models – have been given no opportunities to make decisions. While incentives have been given for market production, the modern technocrats have become the 'knowers' of mass-scale production.

This erosion of decision making has been more crucial to women, small-scale farmers and indigenous groups, who depend most on limited resources – which whilst contributing only marginally to the market economy, have had a greater contribution to make in terms of sustainability. A reversal of this situation will be necessary to empower the people, however challenging it may be.

The prerequisites for Local Action Agenda 21

Under the circumstances discussed above, the prerequisites for Local Action Agenda 21 are:

- changes in development perception
- responsible participation
- changing power relations, and
- changes in the process of development or procedural transition.

Changes in development perception

Development through external incentives and interventions has resulted in neglecting the improvements by local people themselves. From the perspective of improvements in the environment as well as in the lives of people, local knowledge is of direct relevance and needs to be blended with new forms of technically defined systems. The rationales for these new dimensions of development have emerged from the limitation of development avenues that have been adopted in Sri Lanka in the past. The improvements expected by the state, agencies and people are multiple and include improvements in the quality of life and the environment in which people live, interact with and depend upon. There has been an enormous amount of historical evidence showing that indigenous communities have developed locally appropriate strategies in favour of managing resources, and the improvement that has been added by local people to human culture, economy,and local resource management cannot be ignored, although they are often eclipsed by the introduction of new technology, such as miracle seeds. Yet advances that have been made by the people in their immediate ecosystems are numerous and include socio-economic aspects including culture, conservation of soil, water and biological diversity, indigenous knowledge systems and social regulations which respected the community rights to commons.

Organic farming, integrated pest and weed management, multiple use of forestry, multiple use of local flora, water and soil conservation, water resource management, irrigation reservoirs and indigenous water purification are the contributions that have been made by the people themselves for inter-generational sustainability. The exclusion and lack of recognition of such innovations has resulted in a belief that development means externally driven interventions and not the internally evolved, culturally embedded systems of local communities. This suggests that changes are needed in the attitudes of the interventionists as well as in the people themselves through the recognition of local knowledge, and respect for local innovations.

Responsible participation

Agenda 21 proposals call for a community-orientated approach to development and environmental management which requires the full participation of people as partners in the development process. However, attitudes regarding the participation of the grass-roots have been structured by a hierarchical system in which the ability of the people themselves to share the responsibilities has been ignored by policy makers. Under specific circumstances, people have been paid to become responsible for the success of projects. For example, in the 1980s, the state-initiated community forestry project secured people's participation by giving them food aid as part of the programme. Although 'people's participation' has become a buzz-word in development projects, the environmental management measures that have been built into project formulation and which are to be practically adopted by people have never become an integral part of the local resource management.

In Sri Lanka, people's participation in interventional programmes has often been estimated from the numbers that attended the programme, so the people's involvement is evaluated on a head count. In some other cases participation in more practical terms has been secured as a means to accomplish the objectives of the projects. Of course, such programmes have been introduced cheaply, and have been claimed as 'participatory development'. The common feature is that in Sri Lanka, people's participation has been limited to passive attendance and is therefore non-transformative. It is not possible to add different meanings and attitudes with regard to 'responsible participation' of people in Action Agenda 21.

The practical problems impeding responsible participation are numerous. The existing power structure and bureaucracy clearly impede active participation of the poor, especially women, because they do not own or control resources, neither is their social and economic status properly recognised. As has been debated by Nelson and Wright (1995), the officers of external agencies speak the rhetoric of participation yet behave hierarchically. For them participation means attending to project activities on the basis of food assistance. For instance, the reforestation of state lands with the species selected by the Forestry Department on a five-year lease has been noted as participatory. This allows people to grow short durational crops only before the canopy cover comes up, at which time the Forestry Department takes over. These projects do not enable people to select the tree species for planting. In agriculture too the well-adopted ideology is that farmers' knowledge is not transformative and is non-scientific. As a result the experienced farmers, who carry generations of knowledge and experience, lose their confidence in interventions which try to teach them from a base level.

What needs to be incorporated into LA21 is the responsible participation of people, particularly since many of the multi-million, large-scale projects have not been successful in the long run. This is in contrast to those that have been built upon local environmental knowledge and indigenous practices that have been established as part of locally accepted resource management systems, such as the rescue of biodiversity in home gardens (Wickramasinghe, 1994). Ghai and Vivian (1995) have explained that with the threat of environmental degradation in the Third World, the wide range of actions people take at the local level to manage and protect their natural resources has the potential to help reverse, arrest or prevent environmental decline. Measures enforced on people have either failed to be 'internalised', or have never become part of locally appropriate measures, and are acceptable only as long as the 'subsidies' force the adoption. For most of the farmers, and women in particular (Wickramasinghe, 1995a, 1997a) environmental management and the mitigatory measures are integral parts of resource use practices. Their concept regarding resource management is broad, and includes the management of all village ecosystems.

The approaches of outside agencies frequently address the problems of the agencies themselves, rather than those of the rural poor or their environments. To most poor people in rural areas, for whom daily contact with the environment is taken for granted, it is difficult, if not impossible, to separate the management of production from the management of the environment, as both form part of the livelihood strategy of the household or groups (Redclift, 1995). A study conducted in Labunoruwa with the communities in the peripheral areas of Ritigala shows that conservation and livelihood strategies cannot be separated. Biodiversity, soil and water conservation are the practices that have been embedded into non-forest resource management. Household and farm waste are managed as input and as sources of integrated pest and weed management. The commons, forests and farm land are the components of the village ecosystem, in which women's production tasks are performed and conservation measures have been replicated. The local people's innovations are not the examples for peoples' participation, but are locally evolved, people-orientated measures pertaining to sustainable resource management for the twenty-first century. At least the Local Agenda must be able to begin with local initiatives and people's own innovations to make it doable and locally acceptable.

Role and power relations

It is impossible to generalise the roles administrative sectors play on the one hand and people on the other, in the process of development, although they need to be shared in a mutually acknowledged manner. The well-marked hierarchy of state bureaucracy, external agencies and society obstructs communication and interaction. In development projects most top-level positions are held by technically trained men. Women and the poor are listed either as target or beneficiary groups. This situation has resulted in inequitable power relations and an authoritarian approach to intervention. State officers or agencies exercise a privileged position in almost all the projects while the grass-roots, or the real actors, are seen as passive beneficiaries.

A clear demonstration regarding the gap between the state authority and the local community has been recorded in Muriyakadawala. Traditionally, the local community has played a dominant role in maintaining the village irrigation tank known as 'Karamba wewa'. For all the villagers the tank was 'ape' meaning 'ours'. The indigenous community management system was organised under the leadership of 'Welvidane', an experienced farmer selected as the leader. He had to oversee the matters pertaining to paddy cultivation and negotiate with the community members to take decisions on the maintenance of the tank, the area to be cultivated each season, and irrigation matters. In the 1970s, in connection with the agrarian policy reforms, the responsibility for this was placed on a state officer who identifies maintenance needs and reports them to the provincial administration. Although efforts have been made to maintain close contacts with the people, the people themselves believe that the state officer's presence is symbolic of the displaced roles and responsibilities of local communities. The forceful acquisition of local resources has destroyed community resource management even though the state has no capacity to maintain the modern technology. This has resulted in the abandonment of thousands of tanks which have irrigated the dry zone civilization over generations.

This simple example reveals that the acquisition of local resources from the people and the power of authority placed on the state officers has not improved traditional systems. The human and financial cost that the state agencies have incurred as a result of taking over the local systems, sustained over generations, is immense. Although the officer in charge is responsible for overseeing the situation and deciding the area to be cultivated by assessing the amount of water in the tanks, the officers have no capacity or knowledge to maintain the tanks, whilst local people now only have the right to use the resource. The result has been detrimental to the lives of the dry zone farmers. On the part of the government 'the funds for tank rehabilitation' has been a matter negotiated over and over with external donors and lenders. By contrast the previous community innovated strategy was much simpler: maintenance involved desilting the tank and filling the earth bund of the reservoir, whilst the blocking out of the area needing desilting among the farming families, and attending to the tasks on reciprocal basis was overseen by the 'Welvidane' in collaboration with the farmers. This type of 'participatory development' is rare – mainly due to the acquisition of all common resources as state-owned properties. The environmental and development implications of transitions in power relations can produce crisis situations – for example, with regard to large-scale irrigation management, it has been realised that the state has no capacity to support the costly bureaucratic structures that have been created under the Mahaweli Development Programme. Although new interventions are reluctant to call back the community-based indigenous system, a new approach is under investigation in which 'farmer participated management' will use basic hydraulic principles.

Mosse (1995) shows a similar situation in a system of management of tank irrigation systems in South India – where the resource management technology has not been externally forced, but has evolved over generations to meet the water needs and to solve the perceived problems. Further studies in India (e.g. Ambler, 1992) have revealed that the social relations which define the rights, entitlements and obligations have influenced collective action and co-operation. In integrating environmental concerns and development, the lack of community systems and social regulations will encourage external agencies to focus on local capacity building. The situation pertaining to land ownership is diverse. Land is under the control of the state, privately owned large-scale plantations, small-scale farmers and some lease owners. The need for environmental restoration tends to be neglected by the owners because their interest and status differ from the tenants. This implies that empowerment cannot be adopted in isolation from the other two interconnected scenarios – 'responsible participation' and 'equitable opportunities'.

Rather negative impacts both on the lives of local people and their environment have emerged out of state interventions. The crop monoculture production systems promoted in the dry zone areas where slash and burn agriculture has been practised have increased the cost of production and deteriorated the sustainability of the resources. State ownership over the forest has failed to manage them effectively; instead people's unauthorised use tends to continue. What has been lost is a feeling of commitment to manage resources as common properties. The hierarchical roles have led to poor performance in many areas. Prior to the preparation of future action agendas, it is important to identify the roles of external agencies in the development of resources for human welfare. More often external agencies will have to play the roles of facilitators and community mobilisers, rather than advisers, experts and deliverers.

To make a practical change, the need will be to transfer power, introduce a mechanism to promote local control over decisions, actions, and evaluate the impacts. The transfer of power is connected with the transition in the conventional mode of participation, resource distribution, roles and power distribution. According to Chambers (1995) this is a paradigm shift. In the local context this shift has already started since 'participation' has been used as a word pointing to some revolution in the development paradigm. Initially it appeared as a 'cosmetic label' as has been described by Chambers, to please donors more than the local community. A more critical context refers to the transfer of power relations. According to Chambers' analysis it focuses on the empowering process which enables local people to do their own analysis, to take command, to gain confidence, and to make their own decisions. In theory, this means that 'we' participate in their project, not 'they' in 'ours'. In practice, this creates a revolution in the country; if this is to happen, the top-level decision makers will have to work with the people, and orient planning from the grass-roots. This paradigm will be revolutionary – not only for the decision makers and agencies, but also for the people, particularly for women and the poor, currently assigned low status and unable to negotiate with external agencies. The prevailing situation suggests that the social and bureaucratic structure is not in favour of local empowerment – whereby those who have been labelled as beneficiaries are allowed to define their own problems. In fact this is where dominant power relations can be diluted and principles of equitability can be embedded into the process of development.

The local resource management practices (Wickramasinghe, 1997b), particularly forest resource management, reveal that local people have an amazing amount of knowledge, experience and learning to contribute towards Agenda 21. Theoretically it can be argued that it is the grass-roots who manage the local resources with which they have daily contact and who have developed efficient strategies for the betterment of their lives. Chambers (1995) has suggested that determination of priorities in agricultural, forestry, fisheries and other natural resource research should be much more by and through the analysis of experience of local people, weighted to give voice to women, the weak and the poor. The indigenous forest management practices are connected with the principles of community sharing, moral values, spiritual rituals and locally validated resource management practices (Wickramasinghe, 1997b).

Changes in the development process

What needs to be built into development projects are the problems and needs as perceived by the people. The primary goal is not policy-level integration, and a set of recommendations made after a thorough Environmental Impact Assessment, but a 'doable action' for a given geographical location and resource context. The procedural transition for a local action agenda encompasses two aspects: the first refers to a community-orientated approach, a bottom-up process, in place of the top-down. The second refers to a need-orientated procedure, where needs for development can be defined in terms of economic growth and market commodity production, environmental requirements, and people's basic needs such as food, water, shelter, fuel-wood. All these priorities can be addressed through an appropriate planning process, because resources – land in particular – can be managed in an integrated manner to achieve environmental, social and economic goals.

In the past, although in some specific circumstances the word 'participatory' has been prefixed to the project formulation process, development projects have been defined, designed and executed by top-level decision makers. In some circumstances the legislation introduced in favour of conservation has become a disincentive. Legislation pertaining to the harvesting of timber, jak (*Artocarpus heterophyllus*) in particular, is an example. The cumbersome process of getting permits to harvest and transport jak timber has an influence over the farmers who either harvest their trees illegally – losing their self-respect – or go for other species, if the timber needs are crucial.

The disputes and conflicts created by the state agencies are serious because people tend to lose confidence in agencies and act against the law on customary rights that were believed to be correct in their communities. An example of this is the introduction of SALT (Sloping Agricultural Land Technology). *Gliricidia* hedges have been introduced along contours by the farmers of Hapuwala as a conservation measure with a package of subsidies in the lands prone to soil degradation due to tobacco cultivation. It has taken several years to learn from the farmers that locally accepted species are better options than *Gliricidia*, the roots of which hinder crops. Farmers feel that such projects are designed to satisfy external agencies, explaining that 'something external must be introduced to claim for a novelty and technicality': when the subsidies ceased this model was not replicated.

Another critique in favour of procedural transition is in connection with the 'farmers' woodlots' which have attempted to introduce solutions to fuel-wood problems with ADB support. With the help of the World Food Programme, the forestry sector has succeeded in planting *Eucalyptus* on state lands allocated to farmer families on lease. After ten years, however, the goals of this top-down planning have not been achieved: trees remain uncut for fuel-wood, and wood stocks become mature, contributing only marginally to fuel-wood requirements. For a transition to occur, it is necessary to begin with grass-roots experience to ensure that the twenty-first century will head towards sustainable resource management. If land is to be maintained in an ecologically and economically sustainable manner, management decisions must be made by motivated and experienced people. Moreover, there must be co-operation and partnership between local people and policy makers. This implies that the EIA alone will not be able to assess multiple impacts – a consultative process is important to see how people themselves could co-operate from planning to executing a project.

Conclusion

The adoption of Agenda 21 into national policy formulation has been targeted through a process of building up capacities of the environmental agencies in isolation. Whether or not concepts pertaining to sustainable development are well understood, the sectoral development programmes have used the words without realising the necessary prerequisites, particularly the needs for providing equal opportunities. The process has not been decentralised to prepare plans of action to suit the spatially varying local context pertaining to human and environmental needs.

Environmental conservation should be an integral part of resource management of millions of people in the country, yet they are not informed about the national task and the country's commitment to adopt the principles that the country has agreed upon. This implies that, as in the past, the tendency is for a minority in high level positions to formulate action plans, ignoring the implications for the lives of the people in daily contact with environmental resources – specifically women, the poor, farmers and other less privileged sectors of the society. Among many academic discourses, Reardon (1993) has shown that rural women want to play a part in sustaining the natural environment but they cannot embark on schemes which jeopardise their immediate, very precarious livelihoods. Women's priorities have not been accounted for in formulating policies. In the web consisting of three specific strands relating to environment, development and natural resource management, people cannot be excluded. A change cannot be achieved by merely popularising words like participation, joint management and empowerment – the better-off (either in terms of position or otherwise) must be prepared to share power with the under-privileged. What is needed is a transition in people's attitude more than in institutional capacity building. For this purpose, it is necessary to raise public awareness of the objectives of Agenda 21 and the roles of all on equitable grounds.

To accommodate the broader concept, the country has to broaden its vision – accommodating integrated and participative action into the development process. Quite clearly Agenda 21 has brought in a wider concept than that which has been identified in the NEAP. What has been brought to people's attention by Agenda 21 are not merely matters pertaining to environmental conservation, but also the call for changes in the economic development activities of all human beings for a global transition – to reverse environmental degradation, to restore resources and to formulate development activities in an environmentally friendly manner. The final point to be made here is that, in the adoption of mitigatory measures, feasibility and acceptability need re-examining in context, because it is the people who accept or reject suggestions.

The key national strategies must focus on technical and environmental aspects, as well as the most crucial social, cultural, economic and institutional aspects pertaining to sustainability. The EIA is only a requisite to claim that projects have gone through a process to assure they are environmentally non-destructive. The national goals and local priorities must be clearly reflected in LA21, and it will be the means to assure equitable opportunities for the less privileged people in society, particularly for women, rural poor and indigenous people.

References

Ambler, J. (1992) 'Basic elements of an innovative tank rehabilitation programme for sustained productivity', unpublished paper, Ford Foundation, New Delhi.

Beltratti, A., Chichilnisky, G. and Heal, G. (1995) 'Sustainable growth and the Green Golden Rule', in I. Goldin and L. A. Winters (eds), *The Economics of Sustainable Development*, Cambridge: Cambridge University Press.

Carley, M. and Christie, I. (1992) *Managing Sustainable Development*, London: Earthscan.

Chambers, R. (1995) 'Paradigm shifts and the practice of participatory research and development', in N. Nelson and S. Wright (eds), *Power and Participatory Development: Theory and Practice*, London: Intermediate Technology Publications.

Ghai, D. and Vivian, J. M. (eds) (1995) *Grassroots Environmental Action: People's Participation in Sustainable Development*, London: Routledge.

Ministry of Environment and Parliamentary Affairs (1991) *National Environmental Action Plan 1992–1996*, Colombo.

Mosse, D. (1995) 'Local institutions and power: the history and practice of community management of tank irrigation systems in south India', in N. Nelson and S. Wright (eds), *Power and Participatory Development: Theory and Practice*, London: Intermediate Technology Publications.

Natural Resources, Energy and Science Authority of Sri Lanka (1991) *Natural Resources of Sri Lanka*, Colombo: NARESA.

Nelson, N. and Wright, S. (1995) 'Participation and power', in N. Nelson and S. Wright (eds), *Power and Participatory Development: Theory and Practice*, London: Intermediate Technology Publications.

Reardon, G. (ed.) (1993) *Women and the Environment*, Oxfam: UK.

Redclift, M. (1995) 'Sustainable development and popular participation: A framework for analysis', in D. Ghai and J. M. Vivian (eds), *Grassroots Environmental Action: People's Participation in Sustainable Development*, London: Routledge.

Reed, M-A. (1996) 'Research "In the spirit of Agenda 21"', *The 1996 International Sustainable Development Research Conference Proceedings*, West Yorkshire: ERP Environment.

Shiva, V. (1993) 'The impoverishment of the environment: women and children last', in M. Mies and V. Shiva (eds), *Ecofeminism*, London: Zed Books.

Solow, R. M. (1974) 'Intergenerational equity and exhaustible resources', *Review of Economic Studies* 41: 29–45.

Sri Lankan Environmental Journalists Forum (1994) *1993 Citizen's Report on Sri Lanka's Environment and Development*, Nugegoda: SLEJF.

UNCED (1992) *The Global Partnership for Environment and Development: A Guide to Agenda 21*, Geneva: UNCED.

Wickramasinghe, A. (1994) 'Indigenous agroforestry systems: an adoptable strategy for rehabilitation and sustainable management of degraded lands', in P. Panjab Singh, S. Pathak and M. M. Roy (eds), *Agroforestry Systems for Sustainable Land Use*, India: Oxford & IBH.

Wickramasinghe, A. (1995a) 'Understanding environmental changes from the gender perspective', paper presented at the International Geographical Union (IGU) workshop on: Gender, the State and Environmental Changes, 11–15 December, University of Ghana, Legon, Ghana.

Wickramasinghe, A. (1995b) 'The evolution of Kandyan home-gardens: an indigenous strategy for conservation of biodiversity in Sri Lanka', in P. Halladay and D. A. Gilmour (eds), *Conserving Biodiversity Outside Protected Areas: the Role of Traditional Agro-ecosystems*, Gland, Switzerland and Cambridge: IUCN.

Wickramasinghe, A. (1997a) 'Women harmonizing ecosystems for integrity and local sustainability in Sri Lanka', in Anoja Wickramasinghe (ed.), *Land and Forestry, Women's Local Resource-Based Occupations for Sustainable Survival in South Asia*, Sri Lanka: CORRENSA Publications.

Wickramasinghe, A. (1997b) 'Women and minority groups in environmental management', *Journal of Sustainable Development* (1): 11–20.

11 Ground-truthing ecologically sustainable development

Valerie A. Brown

Introduction

There is a growing need for a coherent state-of-the-environment (SoE) reporting system capable of linking decisions made at the global, national and local levels. World wide, uncertainty prevails about the extent of environmental and social improvement or decline resulting from current approaches to ecologically sustainable development. Global conventions, national policies and local programmes are all equally dependent on continual accurate feedback from a sum of their individual localities. Borrowing a phrase from Geographic Information Systems (GIS), there is a need to ground-truth policies and practices by checking them out at the local level. Growing use of the pressure-state-response monitoring framework (see p. 143) by OECD nations and many of their sub-regions and localities has encouraged progress in collecting compatible information, which is useful to decision makers, working mainly from the top down.

Far less progress has been made in the reverse direction: incorporating the results of local environmental monitoring into management decisions at all three levels. While this depressing experience has been reported from a wide range of countries, the response has been to link SoE monitoring to a repertoire of strategies which encourage key stakeholders to apply the results. Strategies for ground-truthing at the local level include an integrative framework for identifying preferred futures, adapted from Canadian Public Health; a European approach to relating local indicators to national and global levels; and a council/community decision-making framework linking indicators of policy, practice and potential impacts to management of local places, developed in Australia.

The power of the local

Local enthusiasm for taking stock of the local environment is not surprising. Local councils are closer to the firing line of environmental disputes than regional or national decision-making systems. At the local level the pressures of human impact on environmental conditions are likely to be directly visible and immediately felt by the community concerned. Local authorities and local industries must solve problems in real time, rather than waiting to complete a budget cycle or a policy development process. Councils' considerable capacity for local monitoring has been further enhanced since 1992. In that year the United Nations Conference on Environment and Development recognised that the role of local authorities in making on-the-spot decisions will ultimately determine the environmental status of the planet (UNCED, 1993).

SoE reporting is routine at the national level for all OECD countries (OECD, 1993a, 1993b). Developing countries such as Fiji and Vietnam are following (Fiji Department of Environment, 1996; Powis, 1996). In parallel with, but seldom co-ordinated with their national reporting processes, local government authorities (LGAs) and local communities have developed their own local environmental monitoring systems (International Council for Local Environmental Initiatives, 1996). In many cases the local monitoring pre-dates national state-of-the-environment reports by some years (OECD Urban Affairs, 1995; Shoalhaven City Council, 1989). While local and national monitoring initiatives maintain their momentum, criticisms mount that this growing body of local and national information on serious pressures on environmental sustainability is not being incorporated into policy or practice at any level (Australian Council for Overseas Aid, 1994; World Bank, 1995).

Agenda 21, the UNCED blueprint for sustainable development for the twenty-first century, puts considerable emphasis on every local authority having a LA21 strategy, that is, reproducing the whole Agenda 21 at the local level (UNCED, 1993). LA21 plans have been adopted widely by local authorities in Europe, the United States and Canada (McLaren, 1996; Whittaker, 1995). The most common form of action towards a LA21 has been the development of local SoE monitoring and local sustainability indicators suitable for use by local authorities and their communities (Local Government Management Board, 1997; OECD Urban Affairs, 1995). An analysis of some of the leading models of good practice in local monitoring still finds a gap between collecting and applying the information (Brugman, 1997).

Australia, with its widespread regulatory requirements for local SoE monitoring, provides a case study of the issues and solutions in incorporating sustainability indicators into mainstream decision making. At one level of analysis it could appear that Australian councils have been much slower to adopt LA21 projects than in other countries (Environs Australia, 1995). On closer examination, the recommendations of Chapter 28 of Agenda 21 have become well established in Australia, but under other labels. These include the council/community partnerships initiated by Healthy Cities and Community Landcare (Thorman, 1997); the information networks established by CouncilNet and RegionNet (Cotter *et al.*, 1994); the intense interest in SoE reporting (Local Sustainability Project, 1997); and the incorporation of environmental management into mainstream council concerns (Adams, 1992; Brown, 1997).

In both Northern and Southern hemispheres, the single strongest aspect of LA21 initiatives has been the spread of local SoE reports (ICLEI, 1996; Local Government Management Board, 1997; Local Sustainability Project, 1997). In the South, New Zealand has national regulatory requirements for all regions to monitor the state of regional environmental resources and incorporate the results into regional management practice, but still identifies a considerable lag between the two. Australia has no such national policy, but all its eight states and territories have some provision for local monitoring by LGAs. This may take the form of the monitoring of particularly vulnerable issues, such as water in Western Australia, or toxic sites in Queensland or it may be more comprehensive. In New South Wales, it is mandatory for all 170 councils to report on the overall state of their natural environment every year. Since 1 January 1998, this reporting must be related to progress towards local goals for ecologically sustainable development.

Taken together, Australian councils are coming close to establishing the capacity for a national network of local monitoring stations. Such a network could allow localities to speak as a sector on community practices supporting environmental sustainability (Australian Local Government Association, 1995; Brown, 1997). If the UNCED goals for LA21 were achieved and every local authority involved world wide, then the sum of the local monitoring provides a place-sensitive global report card. This is not as unrealistic as it may seem. In 1992, governments through the United Nations Conference on Environment and Development, and conservation groups through the International Union for the Conservation of Nature recognised that conservation alone was not enough for a sustainable planet (UNCED, 1993; IUCN, 1992); active human management is required. Global monitoring is not a luxury, but a necessary management tool. The more advanced the remote sensing, satellite monitoring and technical capacity of Geographic Information Systems (GIS), the more information we have of changes in the physical state of the surface of the planet. In the end, this can only be confirmed on the ground at the local level; in GIS language, ground-truthing the information.

Some links between local and global already exist. The OECD framework for Pressure-State-Response environmental reporting has been adopted, with some refinements, by many national and state (provincial) systems (Environmental Systems Program, Harvard University, 1995; OECD, 1994). The pressures (**P**) placed on the environment by the outcomes of human activities affect the state or condition of the natural environment (**S**) and the diagnosis of the impacts on it (**I**) determine the responses made by legislation, economic instruments, professional practices and/or changes in community values (**R**) (see Figure 11.1).

In Australia, the framework is applied nationally, regionally, and by many local authorities as well. The National Local Sustainability Monitoring Forum has agreed to include Impact as a fourth essential dimension of environmental problem-solving (see Figure 11.1). In spite of the progress made in identifying indicators of pressure, state, impact and response, the same fate seems to await Australian national, state and local monitoring systems as was identified in Europe and North America. The invaluable information so collected remains outside management decisions (Department of Local Government NSW, 1997).

National studies of the sources of local environment management information, carried out in 1992 and 1994 and still continuing, have identified some of the barriers to information being incorporated into policy and practice (Brown, 1994;

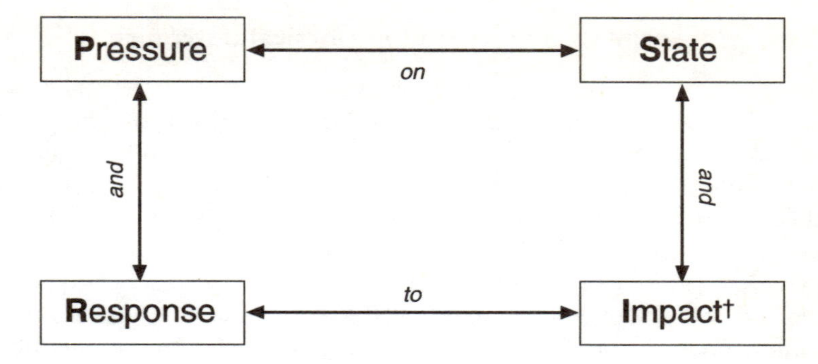

Figure 11.1 P-S-(I)-R* Environmental Monitoring

Notes:
† Impact is included by Harvard University, International Institute for the Urban Environment and Austrailian Local Sustainability Project.
* P-S-R= Pressure-State-Response (OECD)

Brown *et al.*, 1992; Local Sustainability Project, 1997). While extensive expertise has developed in identifying sensitive bio-physical indicators of the state of the environment, there has been no matching increase in ensuring the validity and relevance of the indicators of policies and practices which form the pressure and the response. This has led, in Australia at least, to much political suspicion and distrust of the process itself (Brown *et al.*, 1992). In one closely studied region made up of 21 councils with compulsory SoE reporting processes, only two general managers and very few mayors reported making any use at all of the results (Brown *et al.*, 1997).

These studies of local government authorities, and others on community monitoring (Alexandre *et al.*, 1996) and national SoE reports (SoE Advisory Council, 1996), consistently report the same conditions required for acceptance and application of the information collected by the monitoring process. There is a need to:

- identify the locally preferred future that is to be achieved by ecologically sustainable development;
- monitor the social, economic and bio-physical aspects of each key goal for sustainable environmental management of that locality;
- derive indicators capable of being clearly recognised at local, national and global levels of governance;
- include all key stakeholders directly in the monitoring process; and
- incorporate indicators of the decision-making cycle of policy development, problem solving, and practice into monitoring the sustainability of any given place.

World wide, there are existing initiatives which have satisfied each of these conditions, but separately. Together they offer a blueprint for the monitoring of sustainable policy and practice at the local level in a way that realises its potential to contribute to the validity and reliability of monitoring at the national and the global levels.

Linking local socio-economic and bio-physical outcomes

An integrative framework from public health links sustainable development, social justice and supportive, healthy environments (see Figure 11.2). Applied to environmental management, it requires each environmental issue to be reviewed for its social and economic consequences and vice versa. The integrating concepts are social equity, sustainable development and supportive environments (Dean and Hancock, 1992; Labonte, 1995). The framework has been applied to situations as widely varied as the social, economic and environmental impacts of lead in petrol (Greene *et al.*, 1993); integrated local area planning (Australian Local Government Association, 1995) and municipal health plans (Smith, 1993).

The first step in monitoring local sustainability is the rather obvious one of determining the environmentally sustainable processes to monitor. The basic goals for environmentally sustainable development are intra- and inter-generational equity in the utilisation of natural resources, maintenance of bio-diversity and natural capital, and practise of the precautionary principle of 'do no harm' (World Commission on Environment and Development, 1987). In relation to each of the contributing circles in Figure 11.2, the long-term social goal of social equity, consistent with the cultural values of each society, is intra-generational equity. The long-term economic goal – sustainable development of all resources – requires inter-generational equity. The long-term environmental goal is to maintain a life-supporting environment, locally and globally, which must by definition operate under the precautionary principle. No one of the three can exist independently of the others. The challenge to the monitoring process is to record advances towards all three modes of sustainability, in relation to one another.

This framework for considering integrated management has been applied for over a decade in the international initiative of Healthy Cities (WHO, 1986). Several hundred cities world wide, in developed and developing countries, are implementing the key Healthy City strategies: integrating social, economic and environmental policies; developing supportive environments; strengthening community action; developing skills in negotiation and advocacy; and reorientating services towards prevention rather than cure. The similarity between these goals and those of Chapter 28 of Agenda 21 is striking. A largely untapped experience with monitoring evaluating Healthy Cities initiatives is readily available to local authorities for their environmental management. Examples are Toronto (Healthy Toronto, 1993) and the Healthy Communities Report Card (Hillman, 1996).

Linking local, national and global levels

The first Australian national SoE Report suggests that indicators should be able to be aggregated as much as possible, that is, be amenable to combining with other indicators and to being summed across local and regional levels (SoE Advisory Council, 1996). Yet while bio-physical indicators can be aggregated to a large degree across scales, they have different emphases in, say scattered small colonies of an endangered species as compared to dominant species in a wide range of

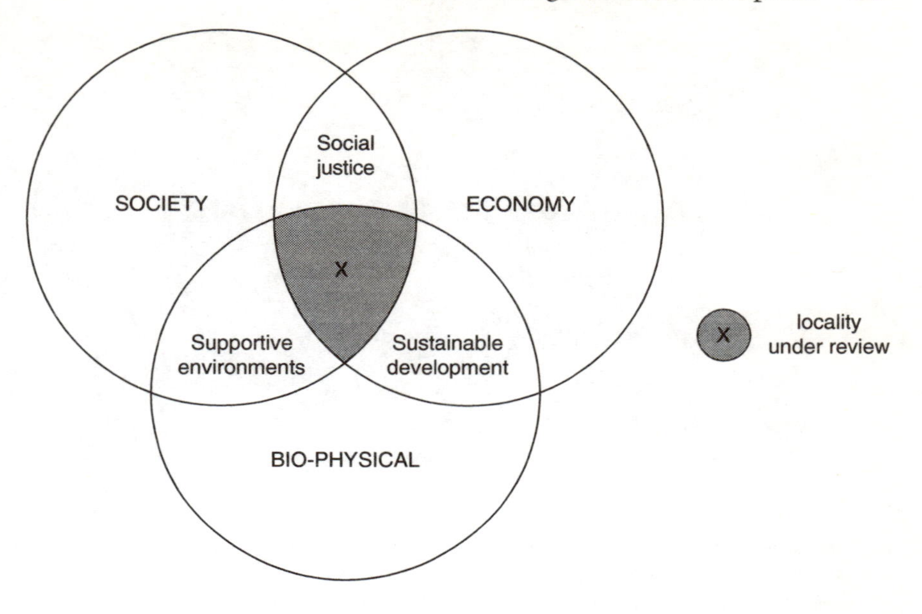

Figure 11.2 Dimensions of sustainable environmental management (Brown 1994, after Dean and Hancock 1992, and Labonte 1995)

ecosystems. Even more disparate are the localised experiences and expectations which make up the priorities of each local government area – the level where national and international policies must in the end be implemented. To attempt to sum social and economic indicators from different places is, in most cases, invalid, and in all cases misleading for decisions at the local level.

One solution to this dilemma is the ABC indicator process which links local, regional and global measures of environmental management (Deelstra, 1994). Trialed by the International Institute for the Urban Environment in 12 European cities, the result is the amoeba model which differentiates between area-specific (**A**), basic set (**B**) and core (**C**) indicators (see Figure 11.3). The themes for the indicators are arrived at first within a recognisable local area, and are to be agreed to by the local politicians, government and non-government agencies, and residents. They are translated into areas of concern through local consultation and further consultation with technical experts. The result is a set of area-specific indicators which can be as rich and varied as the locality itself determines.

Examples of the diversity of council-specific ways for collecting area-specific indicators, and also gaining their agreement on a basic regional set, are described for major European cities by the International Institute for the Urban Environment (Deelstra, 1994), and for over 100 councils Australia-wide by the Local Sustainability Project (Leathem and Brown, 1996). In Australia, Shoalhaven City tied its structural and development plans to its community's vision for the area. The City of the Gold Coast built on a strong technical Geographic

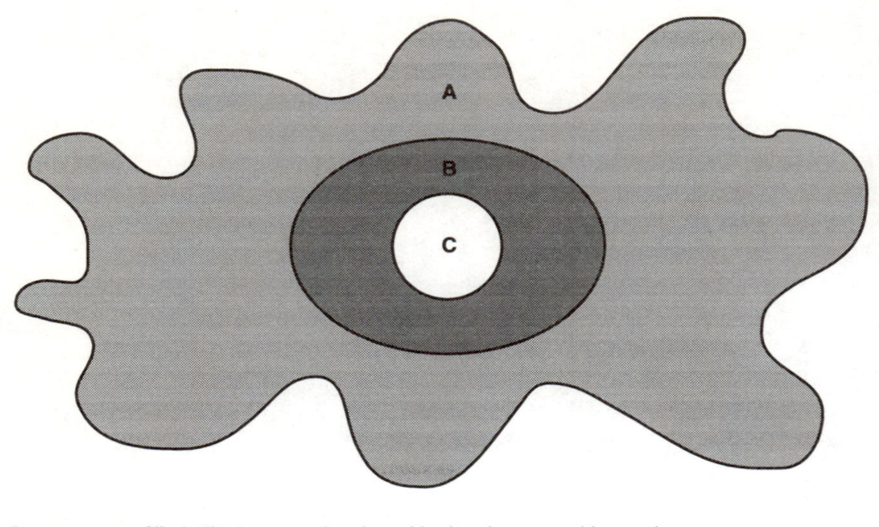

A = area-specific indicators are developed by local communities and authorities to reflect specific local conditions and needs.
B = basic-set indicators are agreed by localities within a region, able to be used for a balanced policy towards sustainable management.
C = core indicators are a small set of indicators common to all regions and essential for national policy development.

Figure 11.3 The A-B-C Indicator Amoeba Model (Deelstra 1994)

Information System base to consult with key decision makers and build council policy in constructing a LA21 plan. Sutherland City Council in Sydney employs a full-time community Ambassador for Local Sustainability to expedite its monitoring process. Yet while recognising the strong individuality of a coastal national park, a high-rise tourist development and the heart of a megalopolis respectively, the indicators arising from the varied monitoring processes matched the key ideas of European cities (see Table 11.1).

After determination of a local Area-specific set of indicators, localities consult within a region to determine the themes in common and the Basic indicator set which can be used across the region. The region itself is defined as a recognised social, economic and ecological entity, and can apply to a catchment, an ecosystem or a nation. In the IIUE project the themes agreed across 12 cities in four European countries are very much the same as those identified almost by every locality. It would be misleading though, to expect that the interpretation would be identical in any two localities in the same region, much less between different countries. A comparison between indicators spontaneously generated across Europe, and in the small coastal resort of Shoalhaven NSW illustrates this point. The two – sophisticated, urban, historic European cities, and a rural and wilderness coastal holiday town – could hardly be more different. Yet their overall needs for local sustainability are highly consistent.

Table 11.1 Self-selected indicators of local sustainability

European cities	*Shoalhaven NSW*
Healthy environment	Keep current landscape in tune with national park
Green space	Balance natural and built environments
Efficient use of resources	Diversify business and industry
Quality of the built environment	Protect identity and character of villages
Accessibility of daily needs	Provide transport for older and younger people
Green economy	Promote tourism which respects our life style
Vitality	Maintain the relaxed friendly atmosphere
Community involvement	Provide a central community facility
Social justice	Meet the special needs of diverse communities
Citizen well-being	Maintain relaxed, friendly stimulating atmosphere

Conditions required for deriving effective Area-specific indicators which could give rise to a basic set of indicators were identified within the European project as:

- a compact set of social, economic and ecological indicators capable of clarifying the process of local sustainability;
- a reference framework recognisable between the regions;
- a distinctly delineated geographic area;
- having identifiable links between local, regional and global policies;
- being easily and directly measurable;
- demonstrating vividly the values being measured;
- connecting to a continuous local evaluation system;
- fulfilling an education function;
- translating between expert knowledge and local experience;
- being compatible with existing monitoring systems.

In Australia, the Australian Local Government Association, the Australian Conservation Foundation and the National SoE Unit in the Department of the Environment have collaborated in a project which has produced three models of good practice of regional monitoring systems (White *et al.*, 1996). This method of producing a shared regional Basic set of indicators is driven from the outside in, that is from the national interest to the local implementation. Another key regional project is arriving at a Basic set of indicators from the inside out. In a region with a strong sense of identity and pride, the Hawkesbury-Nepean Catchment Management Trust (HNCMT) has completed an extensive programme of indicators of the status of natural resources, and is now proceeding to match them to area-specific community priorities before proceeding to identify a basic set (HNCMT 1996).

The international monitoring processes supported by OECD have moved a long way to identifying the third tier of indicators, the Core indicators of the A-B-C model. Indicators of the key global environmental trends such as climate change, loss of bio-diversity, and soil degradation are regularly included in both Area-specific and Basic sets of indicators. This gives opportunities for interpreting ecologically sustainable development for different social and economic systems and status, and supporting locally relevant options for response to the major global pressures on the environment.

Linking policy, problem-solving, practice and place

The established SoE monitoring framework of pressure (P)-state (S)-impact (I)-response (R) works well so far as the separate sections of indicators are concerned. The pressure points (P) of human activities (transport, waste production, agriculture etc.), falling on a given set of bio-physical conditions (S) and will result in particular impacts (I), all of which, once decided upon, can readily be measured. However, the priority-setting mechanisms for the policies which control the pressures, for deciding which impacts are acceptable and which not, and which responses are appropriate, are not so readily assessed. To decide on which themes to measure, and which topics to find indicators for, requires a greater consideration of how the decision-making processes work. The links between P-S-I-R become an important consideration.

This may not be so obvious at the national level, where policies and their expected responses are likely to be designed from a theoretical position and from points of principle. Monitoring results are bulked and the effect of individual decisions and actions are obscured. At the local level the impact of human activities is more immediate and more direct. Individual decisions actually make a difference. The short and the long-term changes impact on the same set of physical conditions and operate within the same set of community priorities. The connections between policy stakeholders, local professional practices and local priorities for action become as important as the individual components, that is, the policies, practices and priorities, themselves.

The four dimensions of any locally based decision-making system: *policy* (P1), *place* (P2), *problem-solving* (P3) and *practice* (P4) support the more generic OECD categories of Pressure, State, Impact and Response (see Figure 11.4). At both local and national levels, the four components together make up an interconnected system which can be entered at any point. The links between the four are never a simple linear relationship but an interactive process. The Local Sustainability Project has applied this as a monitoring framework in partnership with over 50 local government authorities (LGAs), the National Office of Local Government, and the national grass-roots land conservation movement Landcare (Brown, 1997; Brown and Greene, 1994; Local Sustainability Project and Greene, 1996).

Policies (P1) provide the guiding principles and objectives for management decisions and strategic initiatives. In the immediate term they determine local professional and community practice. In the longer term they determine the pressures placed on a particular section of the natural environment by social and economic decisions. Lasting policy directions are not just matters for government or for council executives. Policy is the final agreement by all the stakeholders in a district on the overall direction for the area. In managing for local sustainability, policy development is a response to the condition of the natural resources, social conditions and economic vigour of the place concerned.

In our present system, which separates physical, economic and social policy into three different streams of decision making, it happens only too often that there are three or more conflicting diagnoses of the same issue. At the local level, the three streams are inextricably inter-related. This has been a concern of Australian local authorities for some years, and a major national programme focusing on integrated local area planning (ILAP) was mounted to counteract it (Australian Local Government Association, 1995).

Place (P2) is described by monitoring the bio-physical environment, with the condition of air, living organisms, soil and water now routinely measured. It is less usual to apply methods of monitoring which identify trends and thresholds and so allow predictions to be made. It can also be difficult to gain access to the types of bio-physical information needed for decision making.

Just which particular physical and social issues are selected to work on as a priority will be the result of the local approaches to problem-solving (P3). For instance, drought has been addressed as a crisis intervention for over a century in Australia. Drought relief permitted restocking exhausted soils. In 1995, the national approach changed to accepting drought as an inevitable, routine event in the Australian climate which required preparation and foresight. The principal sustainable management approaches of development, conservation, prevention and rehabilitation were identified as co-existing in every council's policies and practices. Measurements which support the indicators need to take account of the options presented by each one.

Existing practice (P4) may be adequate for implementing a sustainable development policy. It is more likely that, in moving towards sustainability, existing practices are putting pressures on the environment and need to be changed. This applies to professional practice, to individual choices, to industrial activities and to community customs. Currently, where there is maximum progress towards sustainable development, industry and councils are moving to a cleaner production programme, and community groups are adopting sustainable city type initiatives (Local Sustainability Project and Greene, 1996; Smith, 1993). Rural interests have given rise to stewardship of local natural resources through landcare, now also a city movement (Campbell, 1994). Each of these is a major reorientation of current practice.

The 4P framework has been developed in partnership with over 50 Australian councils. It has invariably been effective in identifying shared indicators between the full range of stakeholders, so that those living in the locality have agreed on what is important enough to monitor. The most effective outcome has been when the council/community groups using the framework have worked together over a considerable period of time, have made use of negotiation and conflict management processes, and have arrived at a suite of 12–20 indicators which between them can give an overall evaluation of progress or the need for caution.

Many and varied issues arise in the process of achieving the suites of practical indicators outlined above. Councils can resent the intrusion of community stakeholders into their business. The previous experience of both council and community members can be one of confrontation and lack of trust. The divisions within the community can be strongly entrenched; they may well have existed for decades. The different priorities between the various interest groups may hold deep and very real conflicts of interest, which need to be resolved. Industrial development and jobs, against conservation and environmental sustainability is the classic knee-jerk opposition in every community. The tradition of applying environmental indicators alone to monitor potential solutions increases the appearance of single-mindedness of conservation interests and entrenches opposition.

Shoalhaven NSW held extremely high youth unemployment coupled with a strong real estate lobby anxious to develop the coastal and rural surrounds. The simplistic (and likely) solution was to support commercialisation and so supply jobs, at the expense of the hinterland farms and still undisturbed coastal forests. The 4P process was conducted based on extensive community consultation on a preferred future for Shoalhaven. All possible interests and organisations were involved: fishing, farming, real estate, chamber of commerce, tourist enterprises, conservation groups, primary and secondary school students, churches and service clubs, health services, and the local media, as well as councillors, relevant state services and council planning, community services and environmental management sections. After arriving at a shared preferred future (see Table 11.1), as opposed to a list of individual ambitions, a key strategic group of 48 leading citizens identified the policies, practices, and risks of impact involved in reaching that future. They then proceeded to select a suite of indicators by which all concerned would recognise whether the locality was moving towards or away from their agreed preferred future. This form of monitoring, which enlists media, schools, and public interest organisations as well as council, channels public support towards council staff and local business interests who worked towards the Shoalhaven future; and publicly identifies those who do not.

The preferred future for Shoalhaven included agreeing to maintain the existing attractive rural/coastal/business mix, through limiting building to two-and-a-half storeys and to less than 10 per cent of any settlement or farm. Farmland rates were to be frozen so that farmers would not be penalised by surrounding development; and farms could be multi-homestead to accommodate hobby farmers who did not actually want to farm themselves. An education programme training local school leavers in eco-tourism guiding and small business was to be offered by local interests. The objective was to maximise tourism income through optimising natural attractions, rather than commercialising them. The state government agreed to a sympathetic road by-pass to protect natural attractions, rather than a highway through the main town. Many of these solutions emerged during the consultation process, and were generated by the whole group, rather than sectoral interests.

The following valuable lessons emerged consistently in working with the full range of local stakeholders, together arriving at a sustainable direction for a locality (Griffith *et al.*, 1998):

- Notwithstanding conflicting sets of values and positions on local sustainability in the same locality, there is invariably a strongly held shared vision of the potential of the locality which incorporates inter-generational equity.
- Given a comprehensive goal-setting process, the community vision is always spontaneously described in terms of social, economic and natural environmental outcomes.
- Communities are readily able to distinguish between a wish-list (wants right now) and a long-term vision (changes needed to secure the desired future).
- Communities and government agencies express their goals in different languages which require mutual translation and negotiation.

Lessons from working in this way with councils and their communities on indicators of local sustainability practice are:

- The recognition of the need for different performance indicators for different groups e.g. industry (material flows); government (decisions made and implemented); and community (behaviour changes).
- The need for generic indicators of successful and integrated change management, such as inter-departmental consultation, and establishment of council/community partnerships.

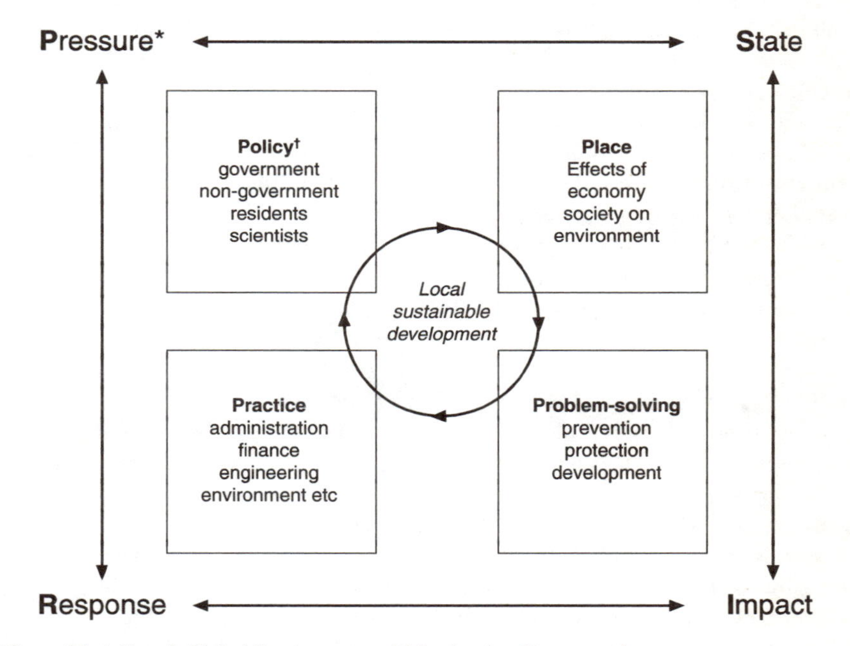

Figure 11.4 Local-Global Environmental Monitoring Framework

Notes:
* PSIR= pressure-state-impact-response (OECD, 1993a; ESPHV, 1984)
† P4 Local decision-making= Policy (P1), Place (P2), Problem-solving (P3) and Practice (P4).

- The essential capacity for monitoring to continue within participants' everyday work programme (specially funded or optional monitoring cannot be sustainable).

Changes in practice will eventually lead in their turn to changes in policy. Titmuss, the father of social policy, defines policy as 'a set of principles directing practice towards predetermined ends' (1974: 23). He suggests that policy development is by definition always concerned with change. If you are not changing anything, you don't need a stated policy, you just go on doing what you have always done. The policy is simply assumed.

Conclusions

The principles of ecologically sustainable development include social and economic equity as well as long-term access to ecological resources. This has been recognised in the Pressure-State-Response monitoring framework used by OECD countries, and now sometimes widened to include identification of the more serious impacts (Figure 11.1). The holosphere of holistic health developed in Canada has been adapted to local sustainable development, identifying the three primary goals for sustainability for which indicators must be sought, namely, supportive environments, social equity and sustainable development (Figure 11.2). Initiatives over the past five years in the Netherlands, the UK, Canada and Australia confirm that community-based identification of their preferred futures will always contain all three components of sustainability.

Councils with their constituent communities provide the base unit for the achievement of local, national and global sustainability. If sustainable development is not practised locally, it does not happen at all. Case studies of localities around the world reveal that a strong support for local monitoring is emerging, creating regional networks documenting the status of supporting resources for local preferred futures. There is an emerging capacity to link these localities together to monitor the state of the planet at the local level. Such a capacity must include the freedom for each locality within its own cultural system to achieve its own goals as well as accept responsibility for contributing to regional and global sustainability. This in turn requires recognition of the gear shift in decision-making systems between area-specific, basic regional and core global sets of indicators of sustainability. A European project monitoring the sustainability of cities established that these three scales of indicators represent a nested set, and not uniform measures which can be simplistically aggregated (Figure 11.3).

In monitoring locally, the objective is to identify the effects of local, regional and national policies and professional practices on local natural resources, when the impact is to achieve sustainable development, with protection of future natural resources (Figure 11.4). An Australian national project has documented the ground-truthing of the outcomes of international conventions, national policies and regional strategies alike through co-operative council/community monitoring in their own locality.

Local sustainability indicators have been found to be capable of reliably monitoring progress towards a shared regional vision of a sustainable future. The three strategies described above join in moving the monitoring of sustainability away from an isolated technical exercise to a tool for mainstream decision making. Their application in a global network of local community monitoring allows for a new level of accountability by governments in implementing international conventions. Acting locally and thinking globally becomes a reality which calls the rhetoric of sustainable development into account.

References

Adams, G. (1992) *Guidelines for the Development of a Local Environment Policy by Local Government*, South Australia: Department of Environment and Land Management.

Alexandre, J., Haffenden, S. and White, T. (1996) *Listening to the Land: Directory of Community Environmental Monitoring Groups in Australia*, Melbourne: Australian Conservation Foundation.

Australian Council for Overseas Aid (ACFOA) (1994) *Summit to Summit: a Community Evaluation of the Outcome of International Conferences 1992–95*, Canberra: Australian Council for Overseas Aid.

Australian Local Government Association (ALGA) (1995) *A Suggested Framework for Preparing Regional Environmental Strategies*, Canberra: Australian Local Government Association.

Brown, V.A. (1994) *Acting Globally: Supporting the Changing Role of Local Government in Integrated Environmental Management*, Canberra: Department of the Environment, Sport and Territories.

Brown, V.A. (ed.) (1997) *Managing for Local Sustainability: Policy, Problem-solving, Practice and Place*, Canberra: National Office of Local Government.

Brown, V. A. and Greene, D. (1994) *Local Government State of the Environment Reports. Summary Report: A Review of the First Year*, Sydney: Department of Local Government and Co-operatives.

Brown V.A., Orr, L. and Smith, D. (1992) *Acting Locally: Meeting the Environmental Information Needs of Local Government*, Canberra: Department of the Arts, Sport, the Environment and Territories.

Brown, V.A., Batros, B., Powell, J. and Williams, R. (1997) *Decision-making Cycles and Local Environmental Monitoring*, Sydney: Hawkesbury-Nepean Catchment Management Trust.

Brugman, J. (1997) 'Is there a message in our measurement? The use of indicators in local sustainable development planning', *Local Environment* 2(I): 59–73.

Campbell, A. (1994) *Landcare: Communities Shaping the Land and the Future*, Sydney: Allen & Unwin.

Cotter, B., Wescott, W. and Williams, S. (eds) (1994) *Managing for the Future: a Local Government Guide*, Canberra: Department of Environment, Sport and Territories.

Dean, K. and Hancock, T. (1992) *Supportive Environments for Health*, World Health Organization Regional Office for Europe.

Deelstra, T. (1994) *Sustainable Cities for Europe*, Delft: International Institute for the Urban Environment (IIUE).

Department of Local Government, NSW (1997) 'Reform of State of Environment

Reporting Provisions of the Local Government Act 1993,' discussion paper, Sydney: Department of Local Government, New South Wales Government.

Environmental Systems Program, Harvard University, (ESPHU) (1995) *Environmental Indicators and Indices*, Boston: Asia-Development Bank and Government of Norway.

Environs Australia (1995) 'Review of Local Agenda 21 Programs in Australia' Melbourne: Newsletter of Environs Australia.

Fiji Department of Environment (1996) *State of the Environment Report*, Suva: Department of Environment.

Greene, D., Berry, M. and Garrard, J. (1993) *Reducing Lead Exposure in Australia. Risk Assessment and Analysis of Economic, Social and Environmental Impacts*, Canberra: National Health and Medical Research Council.

Griffith, R., Brown, V.A. and Ohlin, J. (1998) *Preferred Futures: Sustainable Development for Shoalhaven, NSW*, Canberra: National Office of Local Government.

Hawkesbury Nepean Catchment Management Trust (HNCMT) (1996), *Interim Management Performance Indicators*, Sydney: Hawkesbury Nepean Catchment Management Trust.

Healthy Toronto (1993) *Report on an Index of Sustainability for the City of Toronto*, Toronto: MetroToronto and Healthy Communities.

Hillman, E. (1996) *Signs of Progress, Signs of Caution: a Community Report Card*, Toronto: Healthy Communities Coalition.

Institute of Environmental Studies (ed.) (1996) *Tracking Progress: Linking Environment and Economy Through Indicators and Accounting Systems*, Sydney: Australian Academy of Science Fenner Conference on the Environment, Institute of Environmental Studies, University of New South Wales, 30 Sept. – 3 Oct.

International Council for Local Environmental Initiatives (ICLEI) (1996) *The Local Agenda 21 Planning Guide. An Introduction to Sustainable Development Planning*, Toronto: International Council for Local Environmental Initiatives.

International Union for the Conservation of Nature (IUCN) (1992) *Caring for the Earth*, conservation plan developed at the 18th General Assembly, Perth Western Australia.

Labonte, R. (1995) 'The Holosphere of Health' in C. Chu and R. Simpson (eds), *Ecological Public Health*, Toronto: University of Toronto.

Leatham, J. and Brown, V. (1996) 'Review of 500 Local State-of-the-Environment Reports', Canberra Local Sustainability Project, Centre for Resource and Environmental Studies, Australian National University.

Local Government Management Board (LGMB) (1997) *Local Agenda 21 Project*, London: Local Government Management Board.

Local Sustainability Project (1997) *A Research and Development Program supporting Australian Councils in Local Sustainability Monitoring*, Sydney: Centre for Research in Healthy Futures, University of Western Sydney, Hawkesbury.

Local Sustainability Project and Greene, D. (Consulting Services) (1996) *Getting Ahead of the Game: an Anticipatory Approach to Local Environmental Management*, Canberra: Environment Protection Agency.

McLaren, V. (1996) *Developing Indicators of Urban Sustainability: a Focus on the Canadian Experience*, Toronto: Environment Canada.

Metrocenter YMCA (1995) *Sustainable Seattle, Indicators of Sustainable Community. A Report on Long Term Cultural, Economic and Environmental Health*, Seattle: Metrocenter YMCA.

Municipal Conservation Association (1994) *Managing for the Future: A Local Government*

Guide – Local Agenda 21, Melbourne: Municipal Conservation Association. (Original report written by Deni Green Consulting Services.)

Organisation for Economic Cooperation and Development (OECD) (1993a) *The State of the Environment*, Paris: OECD.

Organisation for Economic Cooperation and Development (OECD) (1993b) *Core Set of Indicators for Environmental Performance Reviews*, Paris: OECD.

Organisation for Economic Cooperation and Development (OECD) (1994) *Selection of Indicators for State of the Environment Monitoring*, Paris: OECD.

Organisation for Economic Cooperation and Development (OECD), Urban Affairs (1995) 'The Application of Social and Environmental Indicators in Local Decision-making', workshop report, Rennes: OECD.

Powis, B. (1996) *Development of an Environmental Health Strategy for Vietnam*, Manila: World Health Organization Report.

Shoalhaven City Council (1989) *State-of-the-Environment Report*, Nowra: Shoalhaven City Council.

Smith, J. (1993) *Review of Municipal Health Plans*, Melbourne: Municipal Association of Victoria.

SoE (State of the Environment) Advisory Council (1996) *National State of the Environment Report*, Canberra: Department of the Environment, Sport and Territories.

Thorman, R. (1997) 'Council and community caring for place', in Brown, V.A. (ed.) *Managing for Local Sustainability: Policy, Problem-solving, Practice and Place,* Canberra: National Office of Local Government.

Titmuss, R. (1974) *Social Policy: an Introduction*, London: Allen & Unwin.

United Nations Conference on Environment and Development (UNCED) (1993) *Local Agenda 21: Chapter 28 of Agenda 21,* New York: Commission for Sustainable Development.

White, T., Alexandre, J. and Thorman, R. (1996) *Regional Environmental Monitoring: Three Case Studies*, Canberra: Australian Local Government Association.

Whittaker, S. (1995) *Case Studies of Local Agenda 21*, London: Local Government Management Training Board.

World Bank (1995) *Monitoring Environmental Progress*, Washington: The World Bank, Environment Department.

World Commission on Environment and Development (1987) *Our Common Future*, Paris: World Commission on Environment and Development.

World Health Organisation (WHO) (1986) *Healthy Cities: Evaluation and Monitoring*, Copenhagen: WHO Europe.

12 Emerging contradictions: sustainable development and the new local governance

Alan Patterson and Kate S. Theobald

Introduction

This chapter examines the contradictions between the ideals associated with Local Agenda 21, and the reality of the ongoing restructuring of local governance in the UK, which has weakened its capacity to respond positively to such a challenge. LA21 includes an explicit demand for democratic participation in the creation and implementation of policies for sustainable development at the local level, and assumes that the elected system of local government will play a leading role. In the UK, however, local government has undergone a process of restructuring which has weakened the role of local authorities and reduced the scope for public participation in decisions about the provision of local services and local policy making in general.

The chapter begins with an examination of the ecocentric concept of sustainability, and contrasts this with the more ambiguous notion of sustainable development, which underpins Agenda 21. Second, the perceived potential role of local government in the development and implementation of sustainable development policies and practices, as part of the LA21 process, is assessed. Third, the chapter confronts the consequences of an ongoing process of restructuring for the ability of local authorities in the UK to incorporate policies for sustainable development into their activities. The rhetoric of LA21 is thus contrasted with the reality of contemporary local government in the UK – a system of local government which has suffered a loss of functions, powers and resources, and a reduction in accountability to local people. In particular, the pernicious impacts of one particular recent form of restructuring – the introduction of compulsory competitive tendering (CCT) – are considered.

Developing policies for sustainable development

The concepts of sustainability and sustainable development lie at the heart of debates about the nature of Agenda 21. Advocates of sustainability, such as Trainer (1995), reject the drive for economic growth which underpins capitalist development, and emphasise the environmental and social benefits of decentralised self-sufficient communities, based upon co-operation and participation, operating within an economic framework radically different to the currently dominant market system. The approach seeks to promote equity between all sectors of society, and between present and future generations, through the rational, and therefore sustainable, use of natural resources.

The idea of economic growth is, however, itself problematic (e.g. Trainer argues that it is actually a 'deeply entrenched myth' (1995: 77)) and the widely used official measure, Gross Domestic Product (GDP) excludes many social and environmental issues, such as income and health inequalities, the pollution of air and water, and access to green open spaces. Indeed, it would be possible to have no economic growth, in the sense measured by GDP, while producing real improvements in terms of equity and the quality of life for the majority of the population. However, Butterfield (1994) points to a serious difficulty in achieving this form of sustainability by noting that redistributive policies coupled with zero growth appear to be unacceptable to the wealthy and politically powerful who would lose out under this scenario. The concept of sustainable development appears to have been developed at least in part to resolve this political impasse. For example, Agyeman and Evans (1994a) argue that the concept of sustainable development can span the divisions between environmentalists, and those economic and political interests that do not have immediately obvious environmental concerns. sustainable development incorporates some aspects of sustainability, such as equity and subsidiarity – the devolution of decision making to the most appropriate level – but, significantly, it does not reject the need for economic growth. Rather, the concept of sustainable development explicitly embodies the assumption that conflicts between environmental protection and continued economic growth can be resolved. However, there is considerable potential for ambiguity in operationalising the concept of sustainable development, and this appears to provide many opportunities for the continuation of unsustainable practices.

International policy development

Whitney (1994) suggests that the 1972 Stockholm Conference (which produced the Stockholm Declaration – identifying the responsibilities of governments to protect and improve the environment for current and future generations) was the starting point of the formal international debate on the environment and development. Later the influential World Commission on Environment and Development (also known as the Brundtland Commission) provided a definition of sustainable development which incorporated the notions of intra-generational and inter-generational equity in the use of the world's resources. The Brundtland Commission's Report (WCED, 1987), which called on the United Nations to establish a Programme of Action for sustainable development, accepted the continuation of a market economy with only minimal regulation, and assumed that economic growth and environmental protection were compatible. O'Keefe and Kirkby (1994) suggest that after 1987 the environmental agenda became increasingly anti-development due to the influence of lobbying organisations such as

Friends of the Earth. However, at the United Nations Conference on Environment and Development (UNCED, widely known as the Rio Earth Summit) held in 1992, the emphasis shifted back in favour of continued economic development. The UNCED has been argued to represent an awareness-raising exercise at the highest political level (Whitney, 1994), the material outcome of which was the documentation of a number of agreements – Agenda 21 being the centrepiece.[1] Agenda 21 was presented as an international programme to achieve sustainable development in the twenty-first century with the aim of protecting the environment through the active involvement of all sectors, agencies and levels of government. However, as Whitney (1994) argues, in reality it is weak and ambiguous precisely because it embodies global compromises acceptable to widely different interests and nations. Thus Agenda 21 does not propose radical policies, instead it assumes that solutions can be found within the framework of a lightly regulated market economy.

Given the caveats outlined above, Agenda 21 seeks to promote sustainable development through the twin ideals of increased public participation and subsidiarity, and particularly emphasises the need for action at the local level – through the establishment of LA21 arrangements (UNCED, 1992: Chapter 28). The document explicitly states that a suitable framework for the co-ordination and management of local environmental action already exists in the form of elected local government.

Central government policy making in the UK

In the UK the government has issued official statements accepting the need for sustainable development since the late 1980s, and elements of sustainable development policies were incorporated within the 1990 White Paper *This Common Inheritance*, the UK's first environmental strategy. The government also established the Central and Local Government Environment Forum which developed Local Government 'Green Charters' covering a broad range of sustainable development issues, and, following the UNCED recommendations, produced its first national strategy for sustainable development, *sustainable development: the UK strategy*, in February 1994. A spokesperson for the Department of the Environment has described a number of other ways in which the government believed progress had been made towards the achievement of sustainable development, including the publication of national indicators, a draft waste strategy, a strategy on air pollution and the introduction of tax differentials to promote environmentally benign practices, for example to encourage the transfer from leaded to unleaded petrol (Hunt, 1995).

However, there are considerable constraints on the implementation of sustainable development policies at the national level in the UK, partly due to the government's decision to continue its policy of relying primarily on market instruments rather than regulation. Breheny (1994) argues that the government's strategy for sustainable development is 'weak', and that this is in line with its belief that both the protection of the environment, and the pursuit of continued economic development, are attainable. Reid (1995) points out that although section 4 of the strategy, 'Putting sustainability into practice', states that all sectors of society have a vital role to play, it shows no awareness of the potential role of central government, in terms of policy making, and also in the provision of guidance, support and resources. The document fails to set out a coherent national strategy for sustainable development but instead provides only a general survey of some of the issues to be taken into account in formulating such a strategy. Although Agenda 21 specifically cites participation as a crucial element in achieving sustainable development, the UK has adopted a top-down approach in which participation is limited to consultation. The government is thus trying to implement an environmental policy framework in which the existing power and decision-making structures are retained.

Reid (1995) argues that the appropriate role for central government should be that of capacity building, enabling and of providing scope for flexibility. The intention of the UNCED was that Agenda 21 be achieved through a 'bottom-up' process, stressing the needs of the poorest, and therefore, as Reid points out, at the national level there needs to be open governance, more information and a better balance between environmental regulation and market mechanisms. He suggests that public participation can only effectively contribute to a process of sustainable development when it is backed by supportive government, which is prepared to empower the population. If bottom-up approaches are to lead to change at higher levels, governments need to devolve power, and give people a real say in decision making about environmental issues. This would require central government to establish an integrated local, regional and national framework which would encourage a co-operative and co-ordinated approach to the implementation of policies for sustainable development. However, in the UK, the system of environmental decision making is non-participatory, and there is a lack of freedom of information and accountability due to the establishment of narrowly focused quangos which do not operate with the same criteria as elected government institutions (see Patterson and Theobald, 1995).

The potential of LA21

Agenda 21 explicitly assumes that the local level is the most appropriate scale at which sustainable development can be implemented and that democratic local authorities are the most effective agents at this level. Agyeman and Evans also emphasise the importance of local action and initiatives:

> It is local action that is likely to develop enduring concern and involvement, and it is local action which will be needed to secure commitment and facilitate democratic control. Moreover, it is 'the local' which can enable experimentation, and permit diversity. Although there must be international, national and regional frameworks and guidance, it is local policy and action which will ultimately deliver sustainability.
>
> (1994b: 198)

Morris (1994) of the Local Government Management Board (LGMB) argues that local authorities see it as their moral responsibility to deliver policies through the LA21 process, and suggests that local authorities see sustainable development as a logical step forward, building on the provision of traditional environmental services. As Baines (1995) argues, local authorities have a major role to play in the implementation of sustainable development policies because they are major players in the local economy, particularly as employers and through their purchasing policies, and because they manage or regulate important aspects of the local environment, and can strongly influence the environmental behaviour of others.

Part of local government's response to LA21 has been to set out a list of requirements for policy making for sustainable development, this includes: partnerships, policy integration, appropriate scale of operation, freedom of information and open government, the precautionary principle, demand management, continuous environmental improvement and the polluter pays principle (Hams, 1994). For these requirements to be met, central government would need to remove restrictions on local authorities in terms of the revenue they are allowed to raise, and permit greater discretion in the use of resources. It would have to give local authorities greater powers of decision making, and provide the resources to enable them to make strategic decisions about environmental management and protection. Such changes would require a radical change in the current central–local relationship, entailing greater power sharing, and a more important role for local authorities. Breheny (1994) also notes the importance of inter-authority co-operation in the development of sustainable policies, and of strategic planning at a regional level. However, it is clear that systems of inter-regional, national and international liaison are also required to ensure that environmental costs are not off-loaded elsewhere.

Geddes (1993) argues in favour of economic integration at the local level, in order to decrease the energy costs involved in long-distance transportation of materials, components, finished commodities, and workers, and he suggests that this should include the development of more locally integrated strategies for public service provision. This could occur through local authorities having the ability to implement purchasing policies which favour environmentally responsible companies, and to encourage local firms to be more environmentally friendly in their working practices. In addition, if local authorities were granted the powers to employ a local workforce, and use locally produced materials, this would enable them to retain employees with a knowledge of the local environment, and would reduce environmental costs associated with the transportation of workers and materials.

Local government restructuring

Until relatively recently, most local public services were provided directly by local authorities employing local people. Local authorities also had the power to decide to use local firms as suppliers of goods, and positively to encourage those firms to introduce environmentally and socially benign practices. Today, following a series of reorganisations which have reduced the functions, powers and resources of local authorities, they have much less potential for either direct action or influence than in the past. This section examines the form of restructuring of local government, paying particular attention to the social and environmental implications of the introduction of compulsory competitive tendering (CCT) for the provision of local public services.

Today, local authorities in the UK are much larger and more impersonal and bureaucratic than in the past. They perform fewer functions and have less autonomy over those that they have retained, and their scope for imaginative policy formation is tightly constrained – both financially and legally. Two powerful legal constraints are the rules of ultra vires and fiduciary duty. The former, developed on an ad hoc basis by the courts and not defined in legislation, is used to limit the scope of local authorities' powers to those areas explicitly specified by parliament, while the latter is used to restrict expenditure, even within these specified areas, to a level agreed between the auditors and the courts to be 'reasonable'. Patterson and Pinch (1995) have described how the elected local state in the UK is being 'hollowed out' by a four-way transfer of functions, power and resources to the centre, to a range of non-elected local and regional agencies (to which may soon be added the proposed new English Regional Development Agencies), to newly developing supra-national systems, and to private firms (see Figure 12.1). This restructuring of the state has resulted in the fragmentation of policy-making and service provision, and a reduction in democratic accountability at the local level.

One of the most recent and most damaging ways in which this fragmentation has been caused has been through the introduction of CCT which requires local authorities to introduce competition and contractual relationships into the public sector by offering private firms the opportunity to tender for the provision of public services. CCT was introduced during the 1980s for a range of local government blue-collar services, including: building and highways construction and maintenance; building cleaning; other cleaning (street cleansing); refuse collection; catering; grounds maintenance; and vehicle maintenance; and was extended to a range of white-collar services through later legislation.

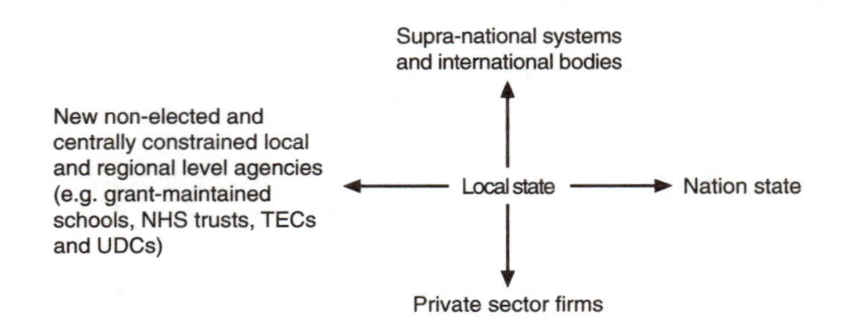

Figure 12.1 'Hollowing out' the local state
Notes:
Arrows represent the flow of functions, power and resources.
Source: adapted from Patterson and Pinch (1995).

The CCT legislation subjects the competitive position of the local authority's own workers, organised into direct service organisations (DSO), to a series of restrictions. It requires that a detailed tender for each of the seven services be advertised by each local authority and, if the authority's own DSO is to be considered for the contract, it must formally submit a written bid. The local authority is prohibited from 'anti-competitive behaviour', which, for example, is deemed to include the packaging of work into large contracts which might deter smaller contractors from competing, and the rejection of lower bids from private contractors without 'good' reason. Moreover, the legislation defines some matters as 'non-commercial' and local authorities are prevented from taking these into account when awarding contracts (this clause principally excludes consideration of contractors' pay rates, terms and conditions, their willingness to incorporate environmentally sensitive practices and materials into their operations, and more general political issues, such as trade links or political affiliations).

The restructuring of local government has caused significant changes in working practices, and the ethos and culture of local government management. There is a new emphasis on contractual relations, for example, in the form of service level agreements between departments, and the client/contractor split established due to CCT, in conjunction with devolved budgets and cost-centre management. This model of contract management is therefore one of decentralised, fragmented decision making and administration, driven primarily by the need to reduce costs. This places serious constraints on the ability of local authorities to implement LA21 (see pp. 164–6). There has been also been a narrowing of the role of management to focus on improving productivity, which is primarily achieved through economies, such as staff cuts, the introduction of performance indicators, and an emphasis on outputs and efficiency rather than on 'citizenship'. This is linked to the focus on the 'customer', which is increasingly allowed to stand as an alternative to electoral accountability (see Sanderson and Foreman, 1996), and is in marked contrast to the principles set out in Agenda 21. There is also an assumption of the transferability of private-sector management principles and practice to local government, yet the relevance of these approaches is questionable, for, as Sanderson and Foreman argue, the 'generic' management model erodes 'public' purpose and local variability/diversity. Local government's political nature also means it has to consider goals and values, and make assessments of need, requiring complex structures that go beyond the profit-orientated focus of the private sector.

CCT has had a fundamental impact on the local government workforce, although these impacts have been neglected within a more recent discourse of concern for the consumers of public services. Much of the competitive advantage of private-sector companies has been derived from their ability to impose worse terms and conditions upon their workers; but, despite this the majority of contracts have been retained by the local authority DSO. This raises the question of the implication for workers within those DSOs of the strategies deployed to maintain services in-house.

There is evidence that CCT has led to the intensification of public service work through cuts in the number of workers (see PSPRU, 1992). The more recent trend has been to cut terms and conditions through increases in basic hours, reduced holiday entitlements and reduced overtime, often in addition to cutting staff (Patterson and Pinch, 1994). Rates of pay have also been reduced in many instances, as the above changes make it easier for managers to take advantage of variations in local labour market conditions and abandon nationally agreed basic rates of pay.

With public service work being increasingly more formalised as a result of CCT, through increased job codification (making work subject to detailed contractual specification) and rule observation (contract and performance monitoring, against details established in job codification) many workers have suffered a considerable loss of autonomy within the labour process. Work is increasingly regimented by contract conditions, for example, reducing the ability of workers to utilise their own discretion about the timing and appropriateness of maintenance work (Patterson and Pinch, 1994). In short, the influence of public service workers is weakened through a formal legal and administrative separation between service delivery and its strategic decision making. More specifically this client/contractor separation combined with the process of competition reduces worker organisation by fragmenting that workforce.

Agyeman and Evans, however, suggest that some good may come of these changes, as a more holistic and integrative approach to policy making is

> beginning to emerge in some authorities as a consequence of the processes of 'market testing' and CCT. These requirements of central government tend to encourage the formation of a more centralised form of local authority administration and decision-making as the work of peripheral departments becomes a potential candidate for some kind of privatisation.
>
> (1994a: 17)

Sadly, this optimism is not confirmed by the results of our research: CCT does not appear to lead to the emergence of a holistic approach to policy making but does lead to the fragmentation of local authority responsibilities, as individual departments lose direct provision of services to the private sector. Rather than encouraging a centralised form of local authority decision making, CCT has meant the devolution of responsibilities and budgets to individual departments, with Direct Service Organisations (DSOs) being expected to operate as stand-alone business units.

CCT has constrained local authorities' ability to give contracts to local firms because they are required, except in exceptional circumstances, to accept the lowest tender for providing a service, and there is accumulating evidence that services provided by local authorities are becoming increasingly concentrated in the hands of a few large national and multi-national companies (Patterson and Pinch, 1995; PSPRU, 1995). There is, as a consequence, a detrimental impact on the environment, in part as a result of the need to cut costs, and, linked to this, the entering of loss-leader bids by large companies which the local authority has to accept, and in part as a direct result of the new forms of contracted-out service provision. The latter is leading to the loss of experienced employees with local knowledge; an increase in journeys due to the transportation of workers and supplies from outside the local area, and the inability of local authorities to influence and put pressure on these private contractors to improve their environmental performance.

Contradictions between CCT and sustainable development

The discussion in this section draws on extensive research into the effects of CCT across the UK as well as the results of an intensive qualitative research project in a cluster of West London local authorities which focused particularly on three traditional local environmental services subject to CCT: refuse collection, street cleansing and grounds maintenance.

Refuse collection requires the deployment of expensive specialist plant and machinery and, as capital outlay on this scale is beyond the capacity of most small companies, consequently private sector-activity in this field is dominated by large multi-national companies and European contractors. Large private contractors appear to have been attracted to refuse collection contracts both because of their high value and their long duration (typically up to seven years) which ensures continuity of work. Street cleansing is also dominated by large companies, although it differs from refuse collection in that the work is a combination of mechanised and manual labour. Some grounds maintenance contracts, such as school playing fields and large public parks are also attractive to private companies because they are of high value and lend themselves easily to highly mechanised and less labour-intensive horticultural work (e.g. grass cutting). In contrast, some specialised horticultural work is more labour intensive and less susceptible to mechanisation.

Our research has shown that there are a number of issues raised by the processes associated with compulsory competitive tendering which combine and inter-relate in terms of their impact on the ability of the local authority to deliver effective environmental services.

Intensification of work and the recomposition of the workforce

The intensification of work is one of the more obvious and well reported of the changes attributable to CCT, and there is a general trend towards staff working longer hours. In one typical local authority, a new contract awarded to the refuse collection in-house team has involved reorganising routes, coupled with longer hours and fewer vehicles, leading to a reduction in the service provided, and an associated impact on the local environment. In grounds maintenance work patterns have also changed, for instance, park keepers have been replaced by mobile patrols, and this has produced a number of detrimental social and environmental effects on the parks and their users.

Local authorities are suffering a loss of skills and experience linked to the recomposition of the workforce, in part due to the intensification of work and the loss of older and more experienced workers who are replaced by younger, often less skilled workers, in addition to a decrease in the training provided. A concern raised by local authorities is that private-sector contractors often target a younger workforce, which the latter believe can be worked hard for a short space of time, then made redundant and replaced. The desire to maximise productivity, linked with cutting costs, has also led to the introduction of 'seasonal hour' arrangements for some workers, for example for the grounds maintenance manual workforce, and has also involved the employment of casual workers on a seasonal basis. Typically, the latter have a lack of local knowledge, and local authority officers readily indicated a number of locally specific ways in which this has a negative impact on the environment.

There is also evidence of a reduction in staff loyalty and commitment (which can obviously impact on service quality) and the general requirement for acceptance of the lowest tenders puts pressure on workers to accept lower pay and a worsening of terms and conditions of employment. The latter coupled to a reduction in job security can lead to a negative impact on local economies.

Contract specifications and the inclusion of environmental criteria

The requirement to subject services to contracting out could be argued to offer some potential advantages because it forces public-sector organisations to rethink the ways in which they provide services, and to address assumptions about the best form of delivering a service. This may or may not lead to a reduction in costs, but could potentially result in an improved service for local people. However, local authorities face annual demands for efficiency savings, and are therefore constantly having to cut costs. Our research shows that the key criteria in awarding contracts is price, and, when linked to a requirement to cut costs, this tends to become a regressive process, as local authorities use CCT as a means of reducing standards. CCT as implemented is, therefore, primarily a mechanism to reduce costs through competition (see Patterson and Pinch, 1995). This process is ongoing, with managers being required to seek cost reductions each time a contract is re-let.

Local authorities are allowed to include certain environmental criteria in contract specifications. They can specify, for example, the types of chemicals and materials to be used, and the frequency of use. However, the scope for including environmental criteria usually depends on the costs involved. The better alternative to environmentally damaging materials may be more costly, and even though some become cheaper in the long term, there are conflicts with the short-term nature of CCT contracts. It has been suggested that the procedure for competitive tendering does enable a contractor's green credentials to be included in the tender evaluation procedure (Hams and Morphet, 1993), however, this does not take into account the wide variations in environmental priorities and standards between local authorities, and the constraints imposed by the requirement that 'specifications do not unintentionally have the effect of restricting or distorting competition' (DoE/LGMB, 1993: 21).

The CCT legislation places the onus on the local authority to prove that the company tendering the cheapest bid is not going to provide the quality required, rather than the contractor having to show how they can provide a service of sufficient quality for the price they have quoted, and the financial situation of local authorities means there is considerable pressure to accept the lowest bid. This reduces the power of the local state to act in support of the local or regional economy by purchasing goods and services from local firms with local purchasing arrangements, which could provide a contribution to environmentally sustainable local economic development (for an elaboration on some of these issues see Pinch and Patterson, 1995).

Flexibility of the contract

Prior to the introduction of CCT considerable flexibility existed in terms of the ability of in-house teams to respond to changes in the demand for local government services. However, under CCT, the flexibility permitted by the contract specification varies, depending on whether the contract has been won in-house or externally, the relationship between the client and the contractor, and the tightness of the specification. In general, the ability to change a specification is limited by the cost of renegotiating the contract, and therefore there has been a reduction in the flexibility to respond to change, and innovation has been stifled. In marked contrast, the workforce in all of the services considered, was expected to be much more flexible.

Monitoring

Some level of monitoring of the contract is required by the local authority to ensure work is done to the specified standard, but this can be expensive, time consuming and conflictual. Monitoring is itself retrospective, and a preferable system would avoid problems in the first place. One possibility is to enlist the end user of the service through a mechanism such as a complaints system, but this possible reliance on the end user could present its own problems – the public are not experts and do not always have the necessary knowledge to know what they should report, and, although there may be some benefits in opening up the process to broader scrutiny, there may also be negative consequences for environmental quality, particularly if there are no general requirements for the contract as a whole to be assessed in terms of its impact on the local and wider environment.

The hollowed out local state

Hardy and Lloyd state that they hold a generally pessimistic view of the scope of local level initiatives for sustainable development, but they go on to argue that:

> many individual local authorities are . . . attempting to integrate the principles of sustainable development via community groups and round table

> discussions. A real case for optimism may well lie at this community level of policy activism rather than at higher levels of responsibility.
>
> (1994: 774)

Their optimism about the involvement of the community (although seen by Agenda 21 as essential to the pursuance of sustainable development at local level) cannot be sustained because the impact of such involvement clearly depends upon local authorities having the power to act upon the results of such consultations with community groups. However, despite these problems we would concur with Wilson, that:

> while the fragmentation of agencies has not helped local authorities to articulate a common community agenda, no other body can easily assume this role, particularly in relation to crucial issues which cross agency responsibilities – like the environment, community safety and health.
>
> (1996: 453)

The currently fragmented forms of local governance cannot provide a coherent policy framework for sustainable development, and this has been exacerbated by the absence of competent democratic forms of regional government which could perhaps act in a co-ordinating role.

The introduction of CCT has, however, undermined the efficacy and capacity of local governance more generally. For example, Stewart (1993) argues that there are severe limitations to government by contract. In particular he discusses the barriers to information flows and other impediments to the organisational learning process which result from the separation of client and contractor. In addition, however, Stewart notes the problems associated with attempts to secure the continuance of a particular set of values in the mode of delivery of a service provided under contract. This in turn chimes with the point made by Nove (1993) that the monitorable targets which have been introduced to check on the work of contractors in the NHS are only indirect indicators of efficiency or service quality and that their value can be undermined as it is difficult to describe any service or product so precisely that a supplier cannot cut corners.

The requirement to specify contracts precisely and to monitor the quality of service provided very closely, together with the need for the new 'cost centres' to send accounts to each other, has resulted in large increases in the number of bureaucrats required to manage the new public sector systems. Nove (1993) notes that this increase in bureaucracy and accounting transactions resulting from the separation of service provision from public responsibility and payment is only one of the costs of introducing contracting as the main means of providing public services.

CCT establishes a separation of a concern for the service to be provided from a concern for the people who will provide that service. Surely here too is an important place for values to be considered. In fact, the CCT legislation specifically excludes a consideration of the pay, terms and conditions of workers providing the service. Second, the CCT legislation also excludes consideration of the activities of the contractor in terms of the stance it adopts on economic, environmental or social issues such as equal opportunities in employment practices, or the maintenance of trading arrangements with apartheid or other unsavoury regimes. Third, the requirement to take the lowest bid (except in exceptional circumstances) reduces the power of the local state to act in support of the local or regional economy by purchasing goods and services from local firms with local purchasing arrangements, which could provide a contribution to environmentally sustainable development (see Pinch and Patterson, 1995). In short, CCT not only contributes to a hollowing out of the local state in terms of its functions, but also of its power to act politically as an employer and as a purchaser of goods and services.

Conclusions

Much hope has been placed in LA21, but it remains ill-defined and in the UK it is being implemented by under-funded and structurally constrained local authorities in the absence of an adequate national or regional framework to co-ordinate the work done at the local level. This chapter has examined the contradictions between the form of public-sector restructuring and the requirements of LA21, and has noted the systematic worsening of the terms and conditions of employment of many local authority workers, and the deterioration in the scope and authority of local government itself. These changes have taken place within a tightly defined commercialised and market-orientated framework which firmly excludes any discourses or frames of reference which embody different value systems.

Given the political impoverishment of the contemporary local government system, local authorities are incapable of successfully implementing a coherent sustainable development strategy. To summarise the earlier discussion which leads to this conclusion: first, there are a number of conceptual obstacles to the creation of a sustainable development strategy, particularly in terms of a lack of understanding at both the national and the local level of how to operationalise the concept. Second, the hierarchical and departmentalised organisation of local authorities has inhibited the move towards a more corporate and holistic approach to environmental policy making. Third, local authorities function within tight legal and financial constraints, for example, legislation on compulsory competitive tendering has severely restricted their ability to provide environmental services to a high standard, and this has occurred at the same time as local authorities are being forced to make significant cuts in expenditure. Fourth, local authorities face considerable structural restrictions, for example, operational limitations due to the size of local authorities, and the absence of a nationally co-ordinated framework for sustainable development, but most significantly, the process of 'hollowing out' local authorities by transferring functions, power and resources to central government, and to a range of non-elected agencies and private firms seriously undermines their capacity to act as competent agents within a LA21 process.

New Labour, new prospects?

The new Labour Government recently signed the European Charter of Local Self-Government which asserts that:

> Local self-government denotes the right and the ability of local authorities, within the limits of the law, to regulate and manage a substantial share of public affairs under their own responsibility and in the interests of the local population.

The previous Conservative Government refused to sign up to this Charter, and there are other signs that Labour may generate a better record on local government than its predecessor. Labour stated in its 1997 general election manifesto that it would give local authorities the powers to set their own council tax and business rate, and would legislate to give local government new powers of community initiative, with a Bill of Rights to individual citizens and to communities. There are also plans for a directly elected mayor for London, and new elected regional assemblies for Scotland and Wales. Unfortunately, in England Labour is intent on creating new non-elected regional level quangos (Regional Development Agencies). Equally unfortunate is the decision not to abolish CCT immediately – instead it looks set to remain in place for at least the next two years until it is replaced with 'Best Value' – a scheme about which there has, as yet, been little specific information other than that it will retain a focus on competition as a means of achieving cost reductions.

There is however currently no commitment to emancipate local authorities by granting powers to introduce long-term policies in line with sustainable development, and there are considerable constraints on the implementation of sustainable development at national and local level in the UK, partly due to the Labour Government's apparent desire to continue with its predecessor's strategy of maintaining a deregulated market economy. The new government also seems willing to acquiesce in the transfer of further power away from the elected forms of the state to business interests. It is about to sign an international treaty, the Multilateral Agreement on Investment (MAI), which will remove the power of national governments to restrict foreign investment (only the defence sector is to be exempted). This will permit the greater penetration of multinational capital into the UK economy, including the provision of local public services. The treaty does permit exemptions to be specified, the USA, for example, has decided to exempt all of its own local and state government from the MAI. Unfortunately the UK government has not been so adroit, and the local government sector will be left open to predatory multi-national activity which will further reduce the capacity of local authorities to pursue the goal of sustainable development.

Note

1 Four other documents were also produced: the Climate Change Convention, the Biodiversity Convention, the Forest Principles and the Rio Declaration.

References

Agyeman, J. and Evans, B. (eds) (1994a) *Local Environmental Policies and Strategies*, Harlow: Longman.

Agyeman, J. and Evans, B. (1994b) 'Making Local Agenda 21 work', *Town and Country Planning* July/August: 197–198.

Baines, C. (1995) 'Local action for sustainability', in Whittaker S. (ed.) *First Steps: Local Agenda 21 in Practice*, London: HMSO.

Breheny, M. (1994) 'Towards sustainable urban development', in Mannion, A. M. and Bowlby S. R. (eds) *Environmental Issues in the 1990s,* Chichester: Wiley & Sons.

Butterfield, J. (1994) 'The New Economics: an ecological approach', in Williams, C. C. and Haughton, G. (eds) *Perspectives Towards Sustainable Development,* Aldershot: Avebury.

Commission of the European Communities (1992) *Towards Sustainability: the Fifth Environmental Action Programme*, Luxembourg: CEC.

Department of the Environment (DoE) (1990) *This Common Inheritance*, London: HMSO.

Department of the Environment and Local Government Management Board (DoE/LGMB) (1993) *The Eco Management and Audit Scheme: A Guide for Local Authorities,* London: HMSO.

Geddes, M. (1993) 'Local strategies for environmentally sustainable economic development', paper presented to the Conference on Urban Change and Conflict, September, University of Sheffield.

Hams, T. and Morphet, J. (1993) 'Hidden peril lurking in the growth of the green agenda', *Municipal Journal* 26 Aug.–2 Sept.: 2–23.

Hams, T. (1994) 'International agenda, local initiative', *Town and Country Planning* 63 (7/8): 204–205.

Hardy, S. and Lloyd, G. (1994) 'An impossible dream? Sustainable regional economic and environmental development', *Regional Studies* 28(8): 773–780.

Hunt, K. (1995) contribution to a workshop on 'National Sustainable Development' at the Annual Conference of the United Nations Environment and Development UK Committee 'Sustaining Development since the Rio Summit', 27 Nov.

Jacobs, M. (1991) *The Green Economy* London: Pluto.

Morris, J. (1994) 'Local Agenda 21 & local authorities' paper presented at seminar, 28 Apr., South Bank University.

Nove, A. (1993) 'Perverse results of a red tape revolution', *Guardian*, 15 Nov., p. 13.

O'Keefe, P. and Kirkby, J. (1994) 'Adding value to nature', paper presented to the 'Struggling with Sustainability' Conference, Staffordshire University, Stoke-on-Trent, 15 Sept.

Patterson, A. and Pinch, P. L. (1994) 'Down with the workers! The impact of CCT on employment in British local government', paper presented to the Annual Conference of the Conference of Socialist Economists, University of Leeds, 9 July.

Patterson, A. and Pinch, P. L. (1995) '"Hollowing out" the local state: compulsory competitive tendering and the restructuring of British public sector services', *Environment and Planning A* 27 (9): 1437–1461.

Patterson, A. and Theobald, K. S. (1995) 'Sustainable development, Agenda 21, and the new local governance in Britain', *Regional Studies* 29(8): 773–778.

Pinch, P. L. and Patterson, A. (1995) 'Public sector restructuring and regional development in the UK' presented at the Regional Studies Association/European Urban and Regional Research Network Conference on Regional Futures, Gothenburg, 7 May.

Public Services Privatisation Research Unit (PSPRU) (1992) *Privatisation: Disaster for Quality*, London: PSPRU.

Public Services Privatisation Research Unit (PSPRU) (1995) *Private Corruption of Public Services*, London: PSPRU.

Reid, D. (1995) *Sustainable Development: an Introductory Guide*, London: Earthscan.

Sanderson, I. and Foreman, A. (1996) 'Towards pluralism and partnership in management development in local government', *Local Government Studies* 22(1): 59–77.

Stewart, J. (1993) 'The limitations of government by contract', *Public Money and Management* 13(3): 7–12.

Trainer, T. (1995) *The Conserver Society: Alternatives for Sustainability*, London: Zed Books.

United Nations Conference on Environment and Development (1992) *Agenda 21*, Geneva: UNCED.

Whitney, D. (1994) 'From the global to the local: issues of policy implementation', in Williams, C. C. and Haughton, G. (eds) *Perspectives Towards Sustainable Environmental Development*, Aldershot: Avebury.

Wilson, D. (1996) 'Structural solutions for local government: an exercise in chasing shadows', *Parliamentary Affairs* 49(3): 441–454.

World Commission on Environment and Development (WCED) (1987) *Our Common Future* (the Brundtland Report), Oxford: Oxford University Press.

13 The opportunities and challenges for local environmental policy and action in the UK

Bob Evans and Susan Percy

Introduction

In this chapter we consider Local Agenda 21 in the UK and examine how the process has unfolded. Our particular concern is to ask whether LA21 may be seen as a fruitful process, likely to deliver a greater local awareness and understanding of sustainable development issues, and an increased capacity to respond to environmental problems, or whether it may be judged as simply another initiative, high profile at present, but inevitably doomed to insignificance and obscurity.

In Britain the response to the call made at Rio in 1992 to think globally and act locally, through the Agenda 21 process, has been taken up by local government in a positive manner and has fired imaginations in a number of local authorities and communities across the UK. The responses have been varied, ranging from, for example, Gloucestershire's Vision 21, led by a voluntary body (Rendezvous Society); Lancashire County Council's LA21 which has been built onto existing State of Environment Reports; and Reading Borough Council with the World Wide Fund for Nature UK's neighbourhood approach called GLOBE – Go Local on a Better Environment.

The Local Government Management Board has recently published an assessment of LA21 in the UK (LGMB, 1997) including 35 case studies which demonstrates the vast amount of activity that has been going on since 1992, including Plymouth's LA21 process focusing on young people, Nottinghamshire's pension fund investment strategy which seeks to link sustainability issues with investment returns, and Cheshire's eco audit.[1] The review suggests that 73 per cent of local authorities in the UK are pursuing LA21s (1997: 15), although there is no clarification as to what pursuing actually means in terms of range or extent of activity. However, the review brings together a large amount of information and examples of LA21s, which are often innovative in scope and wide-ranging in nature.

Given the relatively short time period since the Rio Summit, such progress is remarkable, and deserves to be recognised. LA21 has provided many opportunities for new forms of participation to be developed between the local authority and its community. LA21 has also created the momentum for innovative approaches and new partnerships between the local authority and the community, the local authority and businesses, the community and other organisations. The process has in many cases provided a focus and purpose for local activity and has led to improvements in environmental education and community development within the local community. It has in addition, of course, given local government a rationale for legitimating of its role and purpose.

However, some substantial barriers exist in the LA21 process, which include problems with the rhetoric of Rio and the difficulties associated with translating a contested and complex concept into every-day activities. Related to this, is the difficulty of ensuring that the issues to be explored in LA21 resonate with the concerns of people in their every-day lives. Inertia in administrative structures and reluctance to incorporate LA21 into decision-making responsibilities inhibits LA21 progress, as does political control that is often exerted over the different initiatives stemming from LA21 activities.

Drawing in part upon recently completed research, this chapter outlines the way in which the LA21 process has developed in London and elsewhere in the UK, concentrating on the opportunities and challenges in the process. The chapter offers a critical perspective on the LA21 process and examines some of the contradictions endemic to the process and puts forward some concluding comments on the future prospects of LA21. The comments here are in part derived from recently completed research carried out in South London in which an evaluation of aspects of the LA21 process were examined. Parker and Selman in Chapter 2 of this book point out that 'LA21 programmes are often described as transparent, open and participatory' (p. 19) and our research attempts to examine this claim critically. The study focused upon London boroughs south of the Thames representing a cross section of 'good' and 'bad' authorities in terms of LA21 performance, a variety of political complexions, socio-economic conditions and environmental circumstances. The research was intensive rather than extensive using case studies with the main intention to secure an analysis of the emergence and development of LA21 through an evaluation of the perspectives, expectations and interpretations of the principal actors and agencies involved. The research therefore used semi-structured interviews with key informants: councillors, officers, community representatives, voluntary organisations, environmental activists and the business community, together with the analysis of key publications, committee reports and other relevant publications.

As Selman notes:

> the perpetual dilemma in social research methodology is whether to conduct a quantitative large-scale survey, with the attendant risk of obtaining synoptic but limited information, forced into pre-set categories; or whether to probe a small number of cases which provide rich insights into individual story-lines, but which make it difficult to separate fact from opinion or to generalise from the specific.
>
> (1998: 15)

The following comments have recourse to sources from both approaches, by additionally utilising research conducted by South Bank University for Reading

Borough Council and World Wide Fund for Nature UK on their neighbourhood-based LA21 approach (GLOBE) and on-going research into the national Going For Green Sustainable Communities Pilot Project, together with the evaluation of other secondary sources.

The opportunities

Partnerships

'Real cross sector partnerships are the key to making progress' (Ali Khan *et al.*, 1998: 4). It is perhaps in this area of activity that many LA21s can make a claim for success in establishing partnerships between, for example, the local authority and local community. An example of this is the partnership project involving the local authority, tenants, schools and community organisations in drawing up and implementing environmental improvement schemes for housing estates in the London Borough of Hackney. In Bradford District there are 10,000 small firms and the local authority's Bradford Business and Environmental Support Team (BEST) aims to provide practical environmental support to small businesses. BEST also provides local businesses with information on environmental legislation, encourages environmental action and acts as a platform for the dissemination of best practice information. (LA21 Case Studies, 1997). Voisey *et al.* also point out that 'local authority partnerships with NGOs and local business have the potential to disseminate information, develop shared strategies for LA21 and initiate LA21 projects in areas where local authorities have no explicit role' (1996: 47).

In the Going for Green Sustainable Communities Pilot Project each local authority taking part in the project is partnered with a University which monitors and evaluates the project's progress. Within each of the pilot projects there have been slightly different partnerships formed between the local authority and the associated university, but in all cases there appears to be a supportive relationship and open lines of communication between local authorities, project workers and researchers. 'This has allowed for a great deal of responsiveness to changing needs and conditions on the part of the researchers, and has also avoided methodological strait-jackets' (Smith, J., *et al.*, 1988: 11).

Voisey *et al.* sound a note of caution about partnerships 'however, differences in approach, style, philosophies and objectives form daunting barriers' (1996: 47). In addition partnerships can sometimes become complex and political and be hijacked by dominant interests. It is perhaps too early to say 'whether these arrangements will result in delivering benefits and greater political control over policy to the residents of these areas, but past experience suggests that this will not occur easily' (Deakin and Edwards, 1993, quoted in Aygeman and Evans, 1994: 15).

A frustration that is common to many LA21s is the feeling that the initiative is under-resourced given the nature and extent of the task. Despite the rhetoric about community-based projects and activities, often LA21 has been steered and controlled by the local authority and this has obviously led to a sense of cynicism and distrust in the partnership between the local authority and local community. Evidence from South London suggests that there are feelings of marginalisation and disempowerment and there is a perception that after years of top-down approaches and procedures, people mistrust local authorities' talk of developing partnerships and empowering people. One respondent in South London felt that the 'Community own the process but in practice that is a joke. . . . Many of the resources which will go into producing the LA21 document and the responsibilities for implementing LA21 are the Council's'.

Participation

One of the key aims of LA21 is to encourage participation from all members of society, to improve communications between communities, the local authority and other organisations and thereby increase the involvement of local people in a process through which they can identify their needs and bring these needs into the decision-making arenas. From our research in London it was evident that there were attempts by the local authorities to encourage participation in the process and there have been throughout the UK many instances of new approaches being taken. These include, for example, planning for real, focus groups, parish maps and visioning techniques. One London local authority respondent commented that participation is about 'encouraging others to be involved in the decision-making process' and another said that 'participation is about new mechanisms in bringing people not involved to widen the process of decision making. . . . This is a process that takes many years'. One explanation of why local government is keen to encourage new forms of participation is the self-defence theory in that environmental issues can provide a degree of justification for the support of some locally based services, and also demonstrates the local authorities' enabling role:

> It provides an opportunity to prove to a sceptical local electorate and an unsympathetic central government that they have a useful, popular role in a democratic society. Developing and promoting environmental policies is therefore a way of creating new political space for local authorities through the concept of local guardians of the environment and equally a way of defending their traditional service role.
>
> (Ward, 1993: 466)

While attempts to enable participation are commendable and innovative there is little effective measurement of whether these new methods of engagement are actually reaching the traditionally under-represented members of society such as the elderly, the young, the disabled and ethnic minorities. In many instances those who participate in LA21 are the usual meeting junkies – new methods of participation can still be unrepresentative and do not automatically lead to greater participatory democracy (Selman, 1998).

Selman and Parker (1997) therefore argue for the splicing together of participatory democracy with representative democracy – the latter through local government elections of councillors to represent the views and needs of their electorate and thereby provide the community with a real influence over decision making. The splicing together of these two notions is of fundamental importance, but does require changes in local governance which is likely to be problematic, since, as Selman points out:

> Elected members' attitudes to LA21 tend to vary widely, from feeling threatened, to being excited, to being indifferent. Despite several interesting efforts to incorporate sustainable development into routine committee work, there are still too few systematic attempts to splice representative and participatory modes of democracy in the pursuit of improved local quality of life.
>
> (Selman, 1998: 16)

Local activity and civic pride

LA21 presents local authorities and communities with many opportunities to increase civic pride and focus on local problems and issues. A simple, if somewhat crude example, that many LA21s have supported is the clearance of dog-fouled open space, and whilst this is a simple activity it does help to build up credibility and trust between the LA21 local authority officer and local residents. In all instances it is critical that projects have local relevance and resonate with the concerns and aspirations of the local community. 'Sustainability must be a locally defined and locally acceptable construct, no matter how it is framed at international and national scale' (Voisey *et al.*, 1996: 35). A good example of where this has happened is in Reading, where the neighbourhood-based LA21s have developed a community garden/allotment from a vacant/derelict piece of land, and a local river has been cleaned up. It is interesting to note that:

> there are general themes that seem to be common to all LA21 activity which includes people's concern for transport, particularly congestion and air pollution. Also prominent in terms of common concerns are energy efficiency, recycling and waste minimisation, noise, land use, protection of habitat and open space; again reflecting traditional environmental issues.
>
> (Percy, 1998: 20)

Further interesting issues emerge from the LGMB's review of LA21 progress:

> from the research into the integration of sustainability principles into local authority activities, with the survey data suggesting that the influence of sustainability principles diminishes in areas not classed as traditional environmental issues. Examples include investment strategies, social services, welfare strategies, and anti-poverty measures. This is disappointing but

> perhaps not unexpected given the difficult political, social and economic decisions which are associated with such areas of activity.
>
> (Percy, 1998: 19)

Many of these actions and activities are based on notions of self-help, voluntarism, a sense of community and active citizenship (see Selman and Parker, Chapter 2 in this book for a further discussion of active citizenship). Most activities are carried out by the usual suspects and hence can be viewed as exclusive and unrepresentative. LA21 activity also relies, perhaps rather naively, on the notion of collective community behaviour and action which goes against the political ideology of the last 18 plus years of individualism, privatisation and selfishness.

Energy, imagination and commitment

One of the most positive aspects of LA21 is the vast amount of energy, imagination and commitment that has been shown by many LA21 local authority officers, members and local community representatives. Individuals and small groups of over-worked and over-stretched enthusiasts have been working tirelessly, often with few resources to develop LA21 initiatives. In many cases this has helped to raise the debate and awareness amongst the community of sustainability issues, and has sometimes culminated in locally defined agendas for action. In Reading for example, the neighbourhood groups have developed their own action programmes. The LGMB's LA21 review indicates that 'Some of the case studies are run at arm's length from the local authority, which allows for innovation and imagination but causes problems in getting ideas back into the local authority decision-making structures' (Percy, 1998: 20).

On a note of caution many LA21s seem to be very dependent on the enthusiasm and energy of a key individual/champion, such as the LA21 officer, which has the effect of making the project vulnerable to changes in personnel. It is therefore vital that there is a sense of collective responsibility and commitment for LA21 progress, that this is shared amongst a representative number of people and that local authority officers have appropriate exit strategies in place. It is also important that expectations are not raised unrealistically and that the goals set within the LA21 process are not over-ambitious and hence open to criticism for failure to deliver.

Community development and environmental education

A number of local authorities have realised, as LA21s have evolved, the importance of community development and environmental education as key components in progressing such initiatives. It can be strongly argued that environmental education programmes linked closely to community development provide the means for the delivery of the skills, information and understanding for developing sustainable communities. Financial and time resources need to be devoted to educating for sustainability, which is based on the wider, more embracing, concept of sustainable development rather than the narrowly conceived environmental protection focus that some LA21s are pursuing – so that issues such as a sense of place and poverty are included. Running in parallel to this should be community development work, providing the local community with the skills, knowledge and capacity to deal with the complex and contested issues surrounding sustainability. Unless attention is paid to these facilitating skills and increasing the knowledge base then the following comments, made by an informant in South London, will persist: 'a greater awareness and knowledge of how their [the community's] activities impact upon the environment. . . . The community does not know what Agenda 21 is . . . so what is produced will not be a community document'.

Whilst many local authorities have, rather naively, supposed that communities will come together and develop ideas around the notions of LA21 and that local communities will have a sufficient level of understanding of sustainability, it is evident that this is not the case. Another respondent in South London concluded that 'The community cannot be committed to it (LA21) because they don't know about it. This is a result of a lack of consultation, lack of awareness raising and a need for re-education'. Local authorities that have paid particular attention to community development and environmental education, such as Reading Borough Council, now have self-sustaining LA21 groups working semi-independently on their own LA21s.

Challenges

The previous section demonstrated that there is much to be positive about. The LA21 process, in the UK at least, has shown that there can be new ways of approaching environmental problems and policy, and that these new ways of working can both liberate new energies and provide, often inspirational, solutions to apparently intractable problems. Reflecting this, the Local Government Management Board refers to LA21 as a new paradigm which forces local government to re-evaluate existing practice and to rethink past givens (LGMB, 1997).

However, whilst it is undoubtedly important for organisations and individuals active in the LA21 process to adopt positive attitudes, approaches and vocabulary, it is equally important to recognise the inevitable problems, constraints and limitations on past and future action. This process of evaluation, of looking back as well as ahead, is integral to effective policy formulation and implementation. In this section we consider four of these challenges.

Understanding the concepts

The language of LA21 can be intimidating. Concepts such as 'capacity building' or 'empowerment' are not intuitively understood and yet are central to the rhetoric of Agenda 21. As part of our study in South London we asked respondents to tell us what they understood by these and other concepts commonly employed in the LA21 vocabulary – such as 'participation' and 'partnership'. Our respondents were all active participants in the LA21 process, as local authority officers and councillors or members of local community associations and environmental groups, and yet many had some difficulty in providing a definition of these terms. Although most were able to offer a view on sustainability/ sustainable development (most provided a variant of the Brundtland definition), few knew what was meant by 'capacity building', and interpretations of partnership and empowerment were also hazy.

The environmental debate generally, and the LA21 process in particular, is complex, demanding an understanding of often quite sophisticated ideas. In particular, many people appear to be baffled by the inter-connectedness of processes and functions which typifies environmental policy debates, sometimes including those who are self-selected and active participants in the debates.

The challenge here is not only to make LA21 accessible to a wider cross-section of the population, but also to seek to present environmental and sustainability issues more generally in a form which people can relate to. Most of us refer to our every-day world of work and residence in order to make sense of complex issues, and in this sense LA21 may be the perfect vehicle to encourage global thought and local action. However, on the limited evidence available there is still a need to invest in securing greater knowledge and understanding. Community environmental education will not come cheap, but it is probable that this investment needs to go hand-in-hand with the other elements of LA21.

Sustainability – the critical concept

Sustainability, or sustainable development, is a critical concept in the sense that it is a core notion in LA21, but also because it is clearly a contested concept. The very ambiguity of the term permits widespread acceptance, and makes it acceptable to a wide range of often contradictory social and political interests. Participants in the LA21 process may have clear ideas as to what they are seeking to achieve but, of course these ideas may be mutually contradictory.

At one level this is not a problem. Like justice, freedom and democracy, sustainability may be understood as a social value or *more* which forms part of society's political vocabulary. Although less well established than these other values, it has similarities in that they are all contested concepts, there are many views as to how to secure them, there is a significant gap between the rhetoric and the empirical reality, and like freedom and democracy, sustainability is unlikely to be fully achieved in its pure sense (Baker *et al.*, 1997). In this sense, sustainability is a political rather than a technical term. It is the fact that people *believe* sustainability to be important that is central.

The problem here is that government – central and local – is dealing with a new kind of policy goal which is very difficult to handle. Taking aside the long-term and all-inclusive approach which sustainability policy demands, and which contrasts starkly with the short-term and specific policy approaches which have traditionally characterised government, LA21 and sustainability also require government to dabble in a policy area where there are fewer guidelines as to what might be politically possible and as to what is clearly not. The conventional maps of social class, age, gender, ethnicity and so on which have informed politicians in the past are not so clearly evident in this area of politics.

As sustainable development becomes a more mainstream element of central and local government policy (DETR, 1998), politicians and officials are likely to become more confident about operating in this highly contentious and complex, but nevertheless crucially important area of public policy making.

Opportunism and LA21

Whilst most organisations and individuals participating in the LA21 process are involved because they wish to see changes which will move society towards a more environmentally sustainable future, it also has to be recognised that there are likely to be other, less clearly specified objectives and agendas which can be profitably progressed on the back of LA21. Whilst this opportunism is not confined to environmental policy making, it is necessary to be aware that other motives may stimulate an apparent commitment to the LA21 process.

For example, in our study of South London, it quickly became apparent that one local authority investigated was involved in LA21 because of its wish to be seen as a 'flagship' authority. This objective was perceived as more important than the desire to, say, secure maximum public involvement in LA21 meetings. As a consequence, this particular local authority gave a high priority to LA21 outputs, such as an 'approved' LA21 by December 1996 (the date specified at Rio), but was less concerned to secure some of the more qualitative, and perhaps intangible objectives implicit in the process. Similarly, another local authority studied saw LA21 as an opportunity to support and promote its local economic development strategy, although, as has been noted elsewhere, this may involve support for unsustainable economic growth (Gibbs and Jonas, 1998).

The utilisation of LA21 as a vehicle to secure other objectives is not confined to local authorities. Several commentators have noted that the town planning profession has displayed a high level of interest in the process, not least because the sustainability agenda represents an opportunity for the profession to diversify its interests at a time when its traditional areas of concern have come under pressure from successive governments. It also has to be recognised that individual officers, politicians and academics have identified LA21 as a bandwagon to ride, although in this respect, the policy area is no different from any other.

The defining element of LA21

LA21 is defined, in major part, by its relationship to the community. The distinguishing characteristics of the LA21 approach have been summarised by the International Council for Local Environmental Initiatives (ICLEI) and the United Nations Department for Policy Coordination and sustainable development as follows:

- Multi-sectoral engagement in the planning process, through local stakeholders' groups, which serves as the co-ordination and policy body for preparing a long-term sustainable development action plan.
- Consultation with community groups, NGOs, business, churches, government agencies, professional groups and unions in order to create a shared vision and to identify proposals and priorities for action.
- Participatory assessment of local social, economic and environmental conditions and needs.
- Participatory target-setting through negotiations among key stakeholders in order to achieve the vision and goals set out in the action plan.
- Monitoring and reporting procedures, including local indicators, to track progress and allow participants to hold each other accountable to the action plan.

(ICLEI, 1997)

These key elements convey the central theme of LA21 which distinguishes it from traditional approaches. The community of local businesses, residents, NGOs, churches, etc. are to be active participants in the policy formulation and implementation process as stakeholders. However, whilst this might be a desirable state of affairs, there is little evidence to demonstrate that this is happening, at least in the UK. Both Selman's work (1998) and our own survey in South London indicate very clearly that LA21 exercises have failed to mobilise a new community of stakeholders. In the main, those individuals and bodies who have traditionally participated – usually the educated and articulate middle class – have turned up to meetings, whilst those traditionally excluded – the young, women, members of black and ethnic minority groups – have remained under-represented. In addition, business and commercial interests have shown a reluctance to participate. These approaches are the defining element of LA21, the things which make it what it is. However, in the UK at least, there is little evidence to indicate that much, if any, of this defining element has yet been delivered.

Has LA21 failed to deliver?

The four obstacles briefly outlined above (and of course there are others) serve to illustrate that, despite the publicity, there are still substantial obstacles to be overcome. Whilst much of the literature on LA21 emphasises the positive outcomes of the process and the many examples of good practice, the overall reality is probably less impressive. Whilst the majority of local authorities in the UK are pursuing LA21 (LGMB, 1997), it seems likely that very few are pursuing it at the levels envisaged by the authors of the LA21 document in the pre-Rio Prepcoms. Certainly few, if any, UK local authorities could claim to fulfil ICLEI's key elements outlined above.

So, does this mean that LA21 in Britain has failed? The answer must surely be 'no'. LA21 is at its very heart an ethical process, a set of beliefs marshalled into a policy framework which seeks to operationalise the environmental catechism –

think globally, act locally – by mobilising local communities world wide. In this sense, it cannot fail, since it is clearly a dynamic process. Moreover, there is no evidence to suggest that the ethical imperatives underpinning LA21 are losing political support; on the contrary, it could be argued that the reverse is true, particularly in the UK, where the government has now made a major commitment to the process.

Given that in most countries LA21 has no statutory status, its widespread adoption in such a short period of time is truly remarkable. In part this is because the principles and ideas embodied in LA21 have struck a chord with so many individuals and organisations. The power of LA21 lies in its central claim that new ways of working and new ways of thinking are required in order to deliver environmental and social change, and whilst it may be that this may prove in the long term to be an unrealistic and idealistic claim, it appears that in the UK at least, changes are occurring in local government practices as a result.

Several writers (e.g. Morris, 1998; Selman, 1998) have commented that UK local government has been changed by LA21 and that, as Selman comments, 'there remains a widespread belief that the energy generated by LA21 is unstoppable, and that local government cannot respectably back away' (1998: 17). The degree to which institutional learning has occurred can easily be over-played, but it seems reasonable to assume that, in some local authorities, the experience of LA21 has served to change the ways in which officers and members regard their authority's relationship with the community.

In particular, it may be that there is an increased awareness of the potential contribution to policy making and implementation which may be played by the ordinary citizen through the liberation and utilisation of their local knowledges and understandings – what André Gorz has referred to as vernacular knowledge (1996). This gradual shift away from the primacy of professional expertise can also been seen in the increased interest in devices such as planning for real, round tables and citizens' juries and panels. By the same token, as Parker and Selman argue in Chapter 2 of this book, LA21 has impacted upon the process of citizenship, arguably encouraging more active citizens, and openly espousing the habit of citizenship.

Conclusion

LA21 has not run its course. Although the reality of achievement is probably less than many advocates would claim, it is clear that the LA21 process has secured a range of significant changes in attitude and approach in local environmental policy making in Britain, some more obvious than others. Much of LA21 work is highly commendable and should be celebrated, demonstrating vast amounts of dedication and enthusiasm.

However this energy cannot and will not guarantee the long-term durability of LA21. It is therefore crucial that there is a reassertion that sustainable development initiatives need to be pursued within the context of the local conditions and needs of the area and community, in other words, the recognition of local distinctiveness. It is also important that initiatives resonate with the priorities of local people, and connect to people's every-day concerns and aspirations. Community education and environmental education are key ingredients in the delivery of sustainable life styles, and need to be built in to all sustainable development projects from the start. It is likely that for the longevity of LA21s, legislative and economic incentives will need to be developed hand in hand. Another key element for success is the mechanisms for increasing participation and local democracy, which will inevitably challenge established modes of local authority practice.

We would like to suggest two future directions. First, we wish to caution against the tendency to mainstream LA21. A number of recent publications (DETR, LGA and LGMB, 1998; Morris, 1998) have suggested that the next move for local government should be to establish LA21 as a corporate undertaking, built into everything the local authority does. At one level, this is a plausible argument. The principles of sustainable development do need to be deeply embedded into the process of government, both central and local, and LA21 is the highest profile mechanism currently available for taking this forward. Moreover, given the UK government's new commitment to LA21, and the expectation that all local authorities will have a LA21 process in place by the year 2000, it might be reasonable to see this as the best way to formalise a local commitment to sustainability.

However, we have some reservations over this approach. The essence of LA21 is that it is a *process*, not a *product*. It is a mechanism for establishing agendas, developing local capacities, encouraging and developing understanding and so on. It is not, and should not be owned by any of the participants and stakeholders. Clearly, local government has a pivotal role to play, but local government is not and should not be synonymous with LA21. It is the independence of LA21, its separateness from the established mechanisms of government, which give it its power and authority. Whilst it is totally desirable that local government should seek to implement the spirit of LA21 in policy making and implementation, the mainstreaming of the process, for example through statutory designation, is likely to mean the incorporation of stakeholders to the degree that LA21 becomes simply another consultation mechanism.

Whilst it is important to retain a realistic assessment of what LA21 currently is (in many, if not most UK locations, it is still local authority managed and controlled, and may be legitimately represented as a consultation exercise), it is equally important to retain an awareness of the longer-term objectives for the process as spelled out by ICLEI and quoted above. We would therefore suggest that LA21 should be supported and nutured as an independent and non-statutory process, within which local government has a key supportive role to play. LA21 should be valued as a process which stimulates, offends and provokes. It should encourage independence of mind and build capacity, and it should be an aid to local government decision making and as well as being a perpetual environmental thorn in local government's side.

This leads to our second point. Local environmental policies require local policy delivery mechanisms, and as several commentators have pointed out, such mechanisms do not currently exist in the UK (see e.g. Blowers, 1993; Littlewood and White, 1997). We wish to argue that a clear distinction needs to be made between LA21 as a political, educational and development process on the one hand, and the statutory mechanisms for the delivery of policy on the other. To conflate the two is to consign the potential of LA21 as a force for change to the bureaucratic dustbin.

What is now needed is for the UK government to review the environmental policy delivery mechanisms at its disposal, and to assess how these might best be welded into a national environmental planning system concerned to plan not only land use but all aspects of sustainability policy from energy to waste reduction and disposal. This is precisely the argument put by the Town and Country Planning Association in its call for an integrated, holistic system of environmental planning for sustainability at national, regional and local levels (Blowers, 1993).

Until these effective policy delivery mechanisms are in place, there must be severe doubts over the legitimacy of the LA21 process in Britain.

Note

1 The LA21 Case Studies are issued by the LA21 Case Study project Steering Committee and are available from the sustainable development Unit of the Local Government Management Board.

References

Ali Khan, S., Agyeman, J., and Gibson, M., (1998) *Higher Education 21– Local Agenda 21*, London: Forum for the Future, Department of Transport, Environment and the Regions; Local Government Board and South Bank University.

Baker, S., Kousis, M., Richardson, D. and Young, S. (eds) (1997) *The Politics of sustainable development*, London: Routledge.

Blowers, A. (ed.) (1993) *Planning for a Sustainable Environment*, London: Earthscan.

Deakin, N. and Edwards, J. (1993) 'The enterprise culture and the inner city', in Agyeman, J. and Evans, B. (eds) (1994) *Local Environmental Policies and Strategies*, Essex: Longman.

Department of the Environment, Transport and the Regions, Local Government Association and Local Government Management Board (DETR, LGA and LGMB) (1998) *Sustainable Local Communities for the 21st Century – What and How to Prepare an Effective Local Agenda 21 Strategy*, London: LGMB.

Department of the Environment, Transport and the Regions (DETR) (1998) *Opportunities for Change*, consultation paper on a revised UK strategy for sustainable development, London: DETR.

Gibbs, D., and Jonas, A. (1998) 'Approaching local environmental policymaking: a reconstructed regime perspective', University of Hull Department of Geography Working Paper 98/03.

Gorz, A. (1993) 'Political ecology: Expertocracy versus self-limitation', *New Left Review*, 202, Nov.–Dec.: 55–67.

International Council for Local Environmental Initiatives (ICLEI) (1997) *Local Agenda 21*

Survey: a Study of Responses by Local Authorities and their National and International Associations to Agenda 21, Toronto: ICLEI.

Lafferty, W. and Eckerberg, K. (1997) *From Earth Summit to Local Forum: Studies of Local Agenda 21 in Europe*, Oslo: Prosus.

Littlewood, S. and White, A. (1997) 'A new agenda for governance? Agenda 21 and the prospects for holistic local decision making', *Local Government Studies* 24(4): 111–123.

Local Agenda 21 Case Study Project Steering Committee (1997) *The Local Agenda 21 Case Studies*, London, Sustainable Development Unit of Local Government Board.

Local Government Management Board (LGMB) (1997) *Local Agenda 21 in the UK: the First 5 Years*, London: LGMB.

Morris, J. (1998) 'Coming in from the cold', *Town and Country Planning* 67 (18).

Percy, S. (1998) 'Real progress or optimistic hype?', in *Town and Country Planning* 67 (1): 19–20.

Selman, P. (1996) *Local Sustainability: Managing and Planning Ecologically Sound Places*, London: Paul Chapman.

Selman, P. and Parker, J. (1997) 'Citizenship, civicness and social capital', in *Local Environment* 2 (2) 171–184.

Selman, P. (1998) 'A real local agenda for the century', in *Town and Country Planning* 1: 15–17.

Smith, J., Blake, J., Grove-White, R., Kashefi, E., Madden, S. and Percy, S., (1998) 'Social learning and sustainable communities', unpublished paper.

Voisey, H., Beuermann, C., Suerdrup, L. A. and O'Riardan, T., (1996) 'The political significance of Local Agenda 21: the early stages of some European experience', in *Local Environment* 1(1): 33–50.

Ward, S. (1993) 'Thinking global, acting local? British local authorities and their environmental plans', in *Environmental Politics* 2(3) republished in *Local Environment*, 1997, No 1, 33–50.

14 Constructing future local environmental agendas

Susan Buckingham-Hatfield and Susan Percy

> Environmental problems are *not* problems of our surroundings, but in their origins and through their consequences – are thoroughly *social* problems, *problems of people* their history, their living conditions, their relations to the world and reality, their social, cultural and living conditions . . . At the end of the twentieth century nature is society and society is also '*nature*'.
>
> (Beck, 1992: 81, emphasis in original)

This final chapter attempts to bring together the underlying themes and issues of *Constructing Local Environmental Agendas* and puts forward a framework for future progress in Local Agenda 21. Looking towards the millennium there are many reasons for optimism, but as the book has illustrated, there are also major challenges ahead. This book should not be interpreted as a criticism of what has been achieved but rather as an argument for reform and realignment. We also acknowledge that whilst this book takes an anthropocentric approach to sustainable development, we are not trying to sideline the importance of nature which is a foundation of true sustainability (McLaren *et al.*, 1998).

McLaren *et al.*, in their publication *Tomorrow's World* (1998) put forward a model of the domains of sustainability which shows that sustainability necessarily involves integrating the three conventionally separate domains of economic, social and environmental policy within an overarching political domain (see Figure 14.1). It is within this overarching framework that Figure 14.2 should be considered. We hope that this synthesis will provide a framework for the exploration of issues that we feel is essential if LA21 and sustainable development are to move on in terms of social, technical and institutional changes particularly at the micro level. Thus in the terms of this book we hope that the current barriers within society, politics, economic models and sustainable development goals can be transformed into a source of sustainable local environmental agendas (Irwin, 1995).

As we attempt to bring together the themes that have emerged in the book we offer a framework for sustainable development which is not supposed to be overly prescriptive nor to convey that we think best, since we acknowledge that different situations require different approaches, and it is vital that the local distinctiveness of an area and the character of the community are fully recognised, respected and valued in the LA21 process. Nevertheless, we do strongly believe that unless the

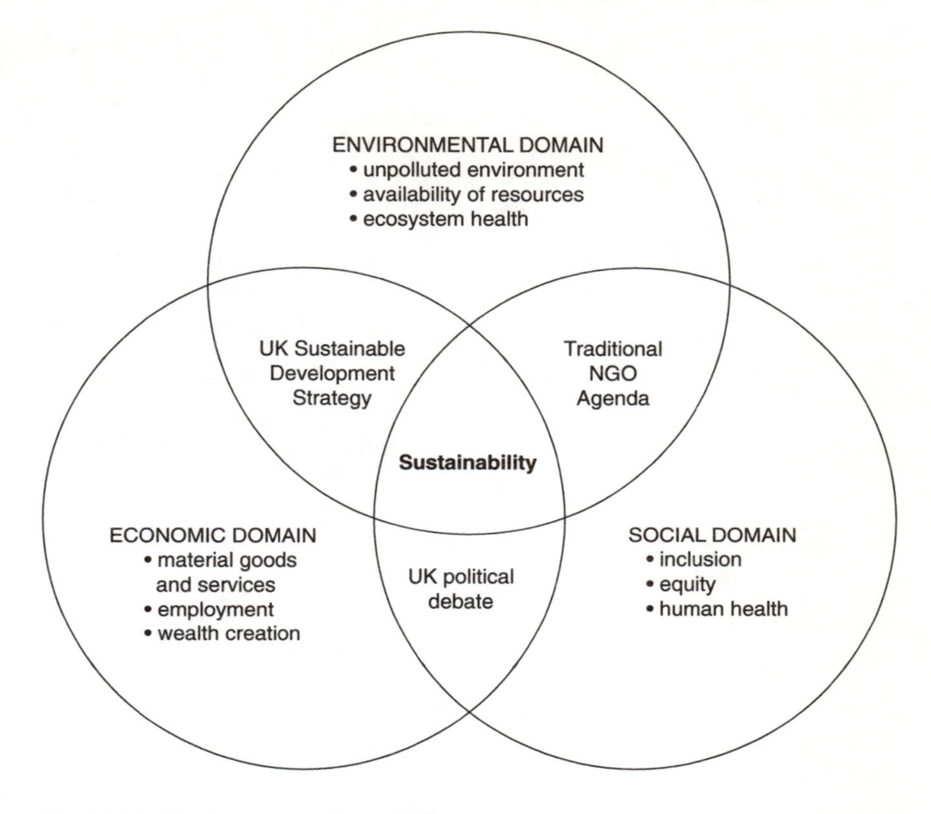

Figure 14.1 The Domains of Sustainability
Source: McLaren *et al.* (1998)

challenges are faced and changes occur, society will find the road to sustainable living winding, bumpy and perhaps ending up in a cul-de-sac.

The challenges are multi-faceted and will inevitably be controversial, some might say unrealistic, and will impact at both the micro level – with individuals needing to take up initiatives – and at the macro level, where collective governmental action will be required. Whilst most of the rest of this chapter focuses on the changes we believe need to be made at the local and national level, it is important to bear in mind that opportunities to effect these changes are severely constrained, particularly in the South, by global forces. The three areas which interact at the local level, which we will discuss in turn, are conceptualised in Figure 14.2: education for sustainability, community development, and local democracy and governance.

Education for sustainability

Although education is provided collectively, it is at the personal level of the learner where transformations are made. Any plethora of education programmes will not

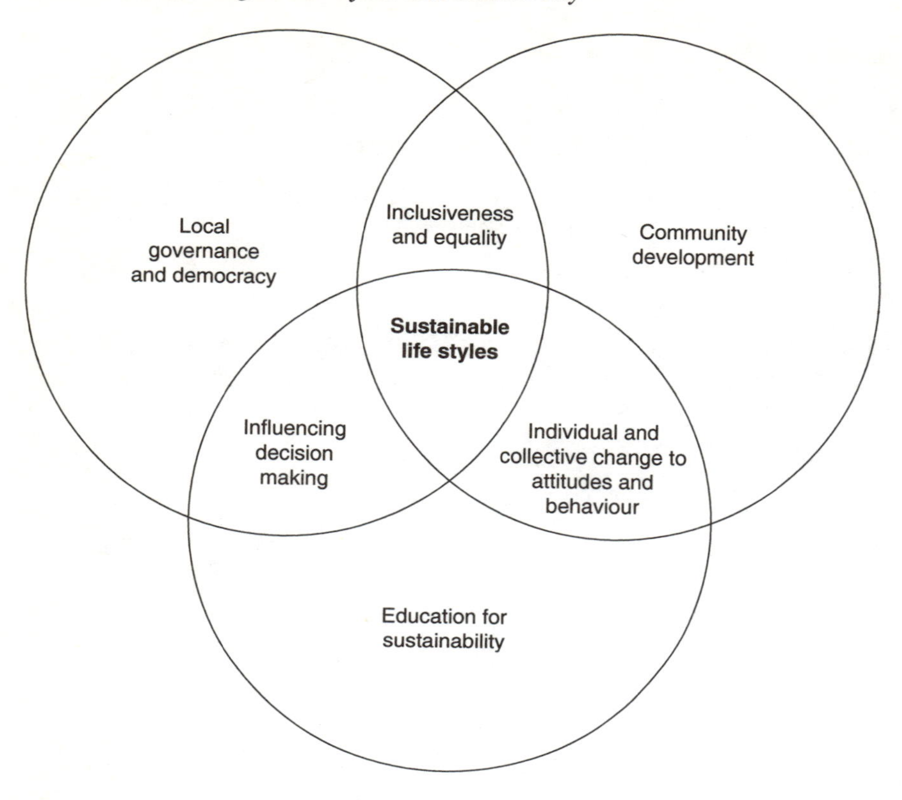

Figure 14.2 The Context for Achieving Sustainable Life Styles

guarantee that people's perceptions and behaviour will be changed. The most successful learning, as any educationalist will testify, takes place when it is the learner who identifies what it is he or she needs to know. Education providers must, therefore respond to this demand rather than prescribe what appropriate body of knowledge should be learned or what outcomes are desired. Failure to do this will result in a small côterie of experts continuing to define the environmental agenda.

The chapters that have preceeded this one have given many examples of individuals and groups of people who want to learn/know more, who want the confidence to apply their knowledge and who want their voices to be heard. Consequently, education for sustainability is about much more than providing environmental knowledge. First, there is environmental literacy – for people to be able to make informed judgements about complex environmental issues they need to be able to understand the science. As Blowers has recently pointed out, the environmental movement is becoming increasingly well versed in what he calls counter-expertise (1997: 166) with which it is able to challenge the status quo. There is clearly a role for schools, colleges, universities and specialist research institutions here. However, formal education needs to be much more outward looking, not only teaching its regular constituency of pupils and students, but making their resources available to the wider community. This can happen through students applying their academic learning to environmental needs in the local community, such as pollution monitoring[1] (see Buckingham-Hatfield, 1992) or by specialist institutes working with local communities: for example, the Royal Botanic Gardens at Kew together with the Brooklyn Botanical Gardens, New York have worked with schoolchildren living in housing projects in New York, teaching them how to grow their own food organically (Royal Botanic Gardens, 1997).

It is not enough, however, to ensure the formal provision of environmental education if this is only reaching a minority of people. As Lubelski and Carmen have pointed out (Chapter 9), only 2 per cent of Pakistan's GDP is spent on education, whilst in many Third World countries, formal education is available only to a relative few. For example, in Pakistan 79 per cent of over 25 year olds have no formal schooling, and 52.7 per cent of the population is illiterate – 78.9 per cent of women. Fifty four per cent of South Asians are illiterate as are 51 per cent of sub-Saharan Africans, 46 per cent of Middle East and North Africans and 24 per cent of East Asians and Pacific islanders (World Bank, 1995). A responsibly acting media also has a role to play in building environmental literacy, as does the internet, but these also have a restricted, privileged, audience at present.

Many of the foregoing chapters give examples of individual and community knowledge being discounted or ignored by policy makers: of children and young people who feel their ideas are not being channelled into decision making (see Freeman, Chapter 6 and Knightsbridge-Randall, Chapter 7), or women whose concerns and unscientifically expressed experience are ignored by politicians (Buckingham-Hatfield and Matthews, Chapter 8) and of agrarian communities whose practised ways of managing the environment are overturned by governments and development agencies eager to implement large-scale farming or energy projects (Wickramasinghe, Chapter 10). Thus, as Wickramasinghe argues, governments, too, need educating in how to utilise local, indigenous knowledge to reverse environmental degradation.

The third role for education for sustainability is to empower people to use their existing knowledge (see Evans and Percy, Chapter 13). In order for any changes to take place which might shift our thinking towards environmental sustainability, we must all believe that our actions can make a difference, that we are capable of creating change, as Leal Filho (Chapter 3) illustrates from both a Northern and a Southern perspective. This is the empowerment that Agyeman and Evans argue can be encouraged through environmental education as a life long process (1994: 233).

Community development

Irwin points out that citizenship in the context of sustainable development 'deals simultaneously with the "public space" and the "private space". The environment, therefore, sits in both public and private space; it challenges us in terms of public policy but also raises personal questions of a profound and ethical nature' (1995: 179).

In order for changes to be made in the public arena, individuals need to be able to take a responsible role in that space. As Clark and Netherwood (Chapter 4) point out, this needs to be more than work (paid or voluntarily) for an NGO. Citizenship in this context requires that we understand each others' needs (of our physically proximate neighbours, but also of fellow human beings world wide) and work with these in the creation of community visions. Parker and Selman (Chapter 2) reinforce the point that a basic prerequisite for meaningful citizenship, however, is a basic standard of living. This issue becomes more acute when countries such as Sri Lanka are considered where, as Wickramasinghe points out, 50 per cent of the population live below the poverty line. Community development, then, needs to encourage informed participation by building the capacity for people to get involved in collective environmental decision making, to provide an effective forum in which people can raise and exchange their concerns in a mutually supportive context and work to overcome poverty and provide a basic standard of living.

Local democracy and governance

Agyeman and Evans strongly argue for the centrality of greater democratisation and involvement of policy making in the advance of sustainability. They state that 'sustainability and the new environmental agenda imply more than simply providing an opportunity for citizen participation in decision making' (1994: 234). This suggests that mechanisms need to be explored that facilitate greater local democracy and which feed into the policy-making machines in government. Community environmental education will provide one vehicle for encouraging local democracy since it 'will not only prepare people for informed participation in the decision making process, but also mechanisms which will enable and assist in building the capacity to actually deliver policy and programmes in partnership with other agencies' (1994: 234).

It has been argued in this book that, at its best, LA21 is opening up decision making to create a space for dialogue between the official and unofficial participants. Parker and Selman suggest that the most durable LA21s exist where lay and expert input are integrated. Simply greater participation on its own, however, will not result in greater democracy – institutions must be changed. Buckingham-Hatfield and Matthews have illustrated in the Australian context, how some changes in administration have appeared to open up the environmental sustainability debate. Both they and Wickramasinghe, writing from different development perspectives, find that governing structures need to be much less hierchical if the dual aims of citizen involvement and sustainability are to be achieved. As Patterson and Theobald (Chapter 12) discuss, the current move towards privatising aspects of government in the UK and elsewhere works against democracy, although it has the (as yet unfulfilled) potential to be used for environmental improvement.

Changes to policy making require a new way of formulating policy that is long term and holistic in nature, so that decision making is inclusive and reflects society's needs. Agyeman and Evans state that there is a 'need for local decision making which is integrated and holistic, and which transcends traditional, professional and departmental boundaries' (1994: 235). Whilst they acknowledge that some local authorities have tried to address this issue with varying degrees of success, they go on to say that there is still no national system of environmental plans which covers the wide range of issues that comes under the umbrella of environmental policy.

Changes beyond the local

At a national level there is a need for commitment of resources and political support for LA21 and sustainable development which will then need to be translated into national strategies for sustainable production and consumption (Carley and Spapens, 1998). This challenge also involves collective governmental action and as Carley and Spapens emphasise the need for the 'orientation of the international development framework – to create more appropriate conditions for development in the south to reduce inequality and initiate the process of sustainable development' (1998: 170). The consequences will include a much more balanced partnership between the North and South.

Equally important is the notion of equality, and the inclusion of marginalised groups. McLaren *et al.* conclude 'it will not be possible to achieve sustainability except in an inclusive society that seeks social justice and well-being for all as its goals'. They go on to explain that 'we also need the political systems to achieve a focus on real well-being and needs. This means the democratic renewal – at local, national and international levels – must lie at the heart of policies for sustainability' (1998: 316). This is the crux of Lubelski and Carmen's argument (Chapter 9) – that LA21, particularly in the South, is in danger of being still born because of all-pervasive powerful market forces.

Despite such forces which threaten to derail LA21 at every turn, it is important to give due consideration to the rhetoric of Rio, since LA21 is about changing every-day life, and not about the arrangements upon which sustainability depends that have been created at international conferences. We need to guard against the inter-governmental discourse becoming a constraint on local initiative, rather than an inspiration. As Lubelski and Carmen argue, there is scope for local people to collaborate across national boundaries, without and even despite the intervention of their governments; their examples show how such collaboration can have mutually enriching results.

What comes through strongly from the contributions to this book, exemplified by Evans and Percy (Chapter 13), is the diversity, commitment and energy at the local level. The local is such an important level at which action and policy will ultimately deliver sustainability (Agyeman and Evans, 1994: 235). Whilst this emphasis on the local is justified, it is also worth highlighting that sustainability will only be delivered if it is linked into changes at international and national levels of governance, together with the other changes discussed above. This implies that constructing local environmental agendas will require the bringing together of the 'social' and the 'natural, the 'local' and the 'global', the 'personal' and the 'public', the 'legal' and the 'voluntary', the 'traditional' and the 'unconventional', and will inevitability lead to changes to institutional arrangements and relations to knowledge and power. These are daunting challenges but we hope that this book at the very least demonstrates that changes are happening. But we all – politicians, policy makers, academics, NGOs, the community and individuals – need to press hard for further changes to help to ensure that in the construction of local environmental agendas the foundations are enduring and long lasting.

Note

1 For example, a Masters student at Brunel University, who is also a school teacher concerned about the impact of noise pollution on children's capacity to learn, has chosen to undertake research into the effects of noise pollution on children's attention patterns, the results of which were used as evidence in the Heathrow Terminal 5 public inquiry. Another student on the same Environmental Change programme (and a primary education specialist) has collaborated with WWF (UK) in researching the usefulness of their environmental education literature.

References

Agyeman J. and Evans B. (eds) (1994) *Local Environmental Policies and Strategies*, Longman, Harlow.

Beck, U. (1992) *Risk Society – Towards a New Modernity*, London: Sage.

Blowers, A. (1997) 'Society and sustainability' in Blowers, A. and Evans, B. (eds) *Town Planning into the 21st Century*, London: Routledge.

Buckingham-Hatfield, S. (1992) *Community Enterprise in Higher Education, a Learning Partnership for the 1990s*, London: CSV Education.

Carley M. and Spapens P. (1998) *Sharing the World. Sustainable Living and Global Equity in the 21st Century*, London: Earthscan.

Irwin, A. (1995) *Citizen Science. A Study of People, Expertise and Sustainable Development*, London: Routledge.

McLaren D., Bullock S. and Yousuf, N. (1998) *Tomorrow's World. Britain's Share in a Sustainable Future*, London: Earthscan.

Royal Botanic Gardens (1997) *Kew*, Autumn issue, Richmond: RBS.

World Bank (1995) *Social Indicators of Development*, Baltimore and London: John Hopkins Press.

Index